CW01558679

About this Book

This book contains a listing of all the service calls undertaken by the lifeboats stationed in Falmouth during 2020.

The inshore lifeboat was called out on 57 occasions and the all-weather lifeboat on 25 occasions. There were several instances where both boats were launched to the same incident and two services when the station's boarding boat was also used.

Both boats also attend quayside services, sailing regattas and Lifeboat days throughout the summer helping to educate and raise awareness of sea safety issues as well to raise funds.

2020 was the year of the Pandemic, which had big impact on the Station and the fundraising undertaken by the volunteers.

All of funds raised by this book go the Falmouth Lifeboat Station in the first instance to meet the costs of our website hosting, software licenses etc.

The costs of the website domain renewals, software, licensing, hosting design and maintenance has been met by a single volunteer for over 25 years since 1996.

For more information please visit our website at:

http://www.falmouthlifeboat.co.uk

Service reports written by Dave Nicoll

Book cover photograph
Falmouth's Severn class all-weather lifeboat *Richard Cox Scott* 17-29 and Atlantic 85 inshore lifeboat *Robina Nixon Chard* B-916 in Falmouth Bay on 18 July 2019. Photo: © Simon Culliford

Table of Contents

RNLB *Richard Cox Scott* 17-29

The Severn class all-weather lifeboat *Richard Cox Scott* entered service at Falmouth on 18 December 2001 and was named by Her Majesty Queen Elizabeth on 1 May 2002 during her Golden Jubilee tour. The Severn class lifeboat is the largest lifeboat in the fleet and is designed to lie afloat and *Richard Cox Scott* is moored alongside the station's pontoon berth allowing 24 hour access at any state of tide. The sweeping sheerline design allows for easier survivor recovery. The propellers and rudders lie in partial tunnels set into the hull that, along with the two bilge keels, provide excellent protection from damage in shallow water.

This lifeboat was funded from the generous bequests of Mrs Ruth Marygold Dix Scott who passed away in May 1998, together with other gifts. Mrs Scott had a love for the sea since her childhood and had lived in Cornwall for many years. The lifeboat is named after her late husband.

Crew: 6/7
Length: 17.3m
Beam: 5.9m
Draught: 1.78m
Displacement: 42 tonnes
Fuel capacity: 5,600 litres
Speed: 25 knots
Range: 250 nautical miles
Construction: Hull: fibre reinforced composite with single-skin section below the chine and 100mm thick foam-cored sandwich above; Deck and superstructure: 25mm foam-cored sandwich.

B class Atlantic 85 B-916 *Robina Nixon Chard*

Robina Nixon Chard was built during 2019 at the RNLI's Inshore Lifeboat Centre at East Cowes on the Isle of Wight with the cost being paid from the legacy of Mrs Robina Nixon Chard.

She arrived in Falmouth by road from Cowes on Thursday 11 July 2019 and was placed on service on 21 August 2019 replacing the Atlantic 75 B-756 *Eve Pank*.

Crew: 3-4
Survivor capacity: 20
Maximum speed: 35 knots, 3 hours maximum
Length: 8.44m
Beam / width: 2.85m
Draught / depth: 0.53m
Displacement / weight: 1.8 tonnes
Fuel capacity: 210 litres
Engines: 2 x Yamaha 4-stroke engines at 115hp each
Construction: Hull: carbon fibre and foam laminate
Structure: includes epoxy glass and foam sandwich layup
Inflatable collar: hypalon-coated nylon

Kayaker in difficulties

Date: 4 January 2020

Location: Nare Head, Gerrans Bay

Weather: SW 1, calm sea with no swell, excellent visibility, clear sky.

Inshore lifeboat: (launched 11.51am / returned 12.44pm). Crew: Luke Wills (Helm), Tamsin Mulcahy, Joe Amps, Lloyd Barron

Details:

Falmouth Atlantic 85 inshore lifeboat *Robina Nixon Chard* launched after a request had been received from Falmouth Coastguard at 11.41am to go to the aid of two kayakers after one had capsized and was unable to get back on his kayak. He had managed to swim ashore and climb on to rocks at the base of Nare Head. The alarm had been raised by the other kayaker via a nearby fishing vessel.

The inshore lifeboat launched at 11.51am and arrived on scene at 12.07pm. The inshore lifeboat crew recovered the casualty from the rocks and after checking that he was alright, they collected his kayak before proceeding to Carne Beach. The inshore lifeboat also escorted the other kayaker back to the beach.

After landing the casualty and his kayak on the beach and ensuring that the other kayaker was also safely back at the beach, the inshore lifeboat returned to station arriving back at 12.44pm.

This was the station's first launch of 2020.

Search for missing person

Date: Friday 10 January 2020

Location: between Bream Cove and Port Navas Creek

Weather: NE 3, slight sea and swell, excellent visibility, clear sky

Inshore lifeboat: (launched 0.46am / returned 3.20am) – Tom Telford (Helmsman), Nick Head, Elliot Holman, Joe Amps

Details:

At 0.38am Falmouth Coastguard requested that Falmouth inshore lifeboat be launched to assist in the search for a missing person from the Mawnan Smith area. The Falmouth & Porthoustock Coastguard Rescue Teams, West Cornwall Search & Rescue Team and a wide range of Police Units including search dogs and the police helicopter were also tasked.

The inshore lifeboat launched from her slipway at 0.46am and arrived on scene at 1.00am to start its shoreline search from Bream Cove to Durgan. On arrival at Durgan the search was extended to Port Navas Creek and then back again to Bream Cove. With the area thoroughly searched at 3.00am the inshore lifeboat was released to return to its station.

The inshore lifeboat was recovered onto its slipway at 3.20am where the lifeboat was refuelled and made ready for service by 3.30am.

Soon after returning to its station the crew of the inshore lifeboat were advised that the casualty had thankfully been located safe and well ashore.

Person in the water

Date: Sunday 19 January 2020

Location: Hemmick Beach, Veryan Bay

Weather: N 1, calm sea and no swell, excellent visibility, sunny

Inshore lifeboat: (launched 1.09pm / Returned 1.26pm) – Jamie Wakefield (Helmsman), Tamara Brookes, Chris Simpson, Caden Harris

Details:

At 1.05pm Falmouth Coastguard requested that Falmouth inshore lifeboat be launched following reports of a person in the water at Hemmick Beach in Veryan Bay. The lifeboat crew were still at the lifeboat station following that mornings training exercise. The Mevagissey Coastguard Rescue Team was also tasked to assist.

The inshore lifeboat launched from her slipway at 1.09pm and having rounded the docks was passing Black Rock when at 1.14pm Falmouth Coastguard confirmed that the person had been recovered from the water so the inshore lifeboat could return to its station.

The inshore lifeboat was recovered onto its slipway at 1.26pm where the lifeboat was refuelled and made ready for service by 1.35pm.

The majority of the lifeboat crew were still at the lifeboat station following that mornings training exercise allowing the inshore lifeboat to be launched in under four minutes.

Person in the water

Date: Tuesday 21 January 2020

Location: Falmouth Inner Harbour

Weather: NW 1, calm sea and no swell, excellent visibility, clear sky

All-weather lifeboat: launched 9.28pm / returned 10.17pm) – Jonathon Blakeston (Coxswain), Dave Nicoll, Tom Bird, Jonathan Hackwell, Sandy Procter, Adam West, Andy Edwards

Inshore lifeboat: launched 7.55pm / Returned 10.25pm) – Neil Capper (Helmsman), Tamara Brookes, Chris Simpson, Caden Harris

Boarding boat: (launched 8.15pm / Returned 10.25pm) – Elliot Holman (Helmsman), Tamsin Mulcahy

Details:

At 7.48pm Falmouth Coastguard requested that Falmouth inshore lifeboat be launched following a report of a punt adrift off Port Pendennis Marina in Falmouth Inner Harbour. The boat was showing signs of recent occupation. The Falmouth Coastguard Rescue Team was also tasked to assist with the search.

The inshore lifeboat launched from her slipway at 7.55pm and headed to Port Pendennis Marina. Having inspected the punt, the inshore lifeboat commenced a search of the Inner Harbour at 7.59pm while the Coastguard teams searched the shoreline.

At 8.08pm the inshore lifeboat requested that the stations boarding boat be launched to assist with the search. The boarding boat launched from its berth at 8.15pm and joined the search at 8.18pm.

After an hour of searching and with nothing being found the decision was made to extend the search to cover the docks and entrance to the harbour so the all-weather lifeboat and the Coastguard helicopter were requested at 9.18pm. The all-weather lifeboat left its pontoon berth at 9.28pm and commenced its search of the moorings between Trefusis Point and Flushing while the helicopter searched the harbour entrance using its FLIR camera. Once the lifeboats had completed their search of the harbour, they were requested to carry out a search of the dock's wharfs.

With nothing being found the Coastguard terminated the search, releasing the all-weather lifeboat at 10.11pm and the inshore lifeboat and boarding boat at 10.20pm to return to their station.

The all-weather lifeboat was back alongside its pontoon berth at 10.17pm, the boarding boat was back on its berth at 10.25pm and the inshore lifeboat was recovered onto its slipway by 10.25pm. All boats were refuelled and made ready for service by 10.40pm.

The initial indications were that the punt had recently been occupied as its engine appeared to be warm and its oars were rigged but despite an extensive search no sign of its occupant was found, and no one was reported missing.

Medical Evacuation from Merchant Tanker Seaturbot

Date: Saturday 8 February 2020

Location: 3 miles South of Pendennis Point

Weather: SW 7, moderate sea and swell, good visibility, overcast

All-weather lifeboat: (launched 2.10pm / returned 3.02pm) – Jonathon Blakeston (Coxswain), Dave Nicoll, Andy Jenkin, Tamara Brookes, Chris Simpson, Caden Harris, Jemima Henstridge-Blows

Details:

At 2.00pm Falmouth Coastguard requested that Falmouth all-weather lifeboat be launched to carry out the medical evacuation of the master of the German Merchant Tanker Seaturbot in Falmouth Bay.

The all-weather lifeboat left its pontoon berth at 2.10pm and headed out into Falmouth Bay. The tanker had anchored in a position 3 miles South of Pendennis Point and was using its engine to provide a lee for the lifeboat. The lifeboat arrived on scene at 2.30pm and having gone alongside the vessels starboard side the casualty who was able to walk made his way down the ships ladder to the lifeboat. Once safely on board at 2.30pm the lifeboat headed back towards Falmouth.

The all-weather lifeboat arrived back alongside its temporary pontoon berth at Port Pendennis Marina at 3.02pm. The casualty was taken to the lifeboat station before being transported to Treliske Hospital. The lifeboat was made ready for service again by 3.15pm.

The 178m (21,353 GRT) tanker had been transiting the English Channel when its master had been taken ill, so the vessel had radioed the coastguard for medical advice. The shore-based doctor advised that the master should be landed ashore for treatment, so the ship diverted to Falmouth Bay where the coastguard arranged for the lifeboat to meet it.

Report of Paddle Boarder in difficulties

Date: Saturday 15 February 2020

Location: off Middle Point, Carrick Roads

Weather: SSW 8, rough sea and moderate swell, poor visibility, overcast with heavy rain

All-weather lifeboat: (launched 8.30am / returned 9.15am) – Luke Wills (Second Coxswain), Dave Nicoll, Andy Jenkin, Elliot Holman, Jamie Connolly, Chris Simpson, Lloyd Barron

Inshore lifeboat: launched 8.17am / Returned 9.10am) – Sandy Procter (Helmsman), Jamie Wakefield, Tamara Brookes, Tamsin Mulcahy

Details:

At 8.10am Falmouth Coastguard requested that Falmouth inshore lifeboat be launched following a report of a paddle boarder who was believed to be struggling in adverse weather conditions off Middle Point in the Carrick Roads. The Falmouth Coastguard Rescue Team was also tasked to assist with the search.

The inshore lifeboat launched from her slipway at 8.17am and arrived off Middle Point at 8.22am, being tasked with searching the Carrick Roads. Given the gale force conditions the inshore lifeboat requested at 8.26am that the all-weather lifeboat be launched to assist in the search.

The all-weather lifeboat left its pontoon berth at 8.30am and headed up the Carrick Roads commencing its search at 8.37am.

Shortly afterwards the inshore lifeboat located a paddle boarder near the Carrick Buoy who said that they were escorting their son but had lost contact with him. The inshore lifeboat continued its search locating the missing son near the Pill Buoy. Once it had been confirmed that there was no one else missing and the two paddle boarders had reached Turnaware Bar they were left to continue upriver while at 8.55am both lifeboats were released to return to their station.

The inshore lifeboat was recovered onto its slipway by 9.10am, while the all-weather lifeboat was back alongside its pontoon berth by 9.15am. Both boats were refuelled and made ready for service by 9.30am.

The father and son had been on a training paddle for a future race and had launched from Gyllyngvase Beach with the intention of paddling up to Truro. While they were well experienced, given the adverse weather conditions people had been rightly concerned for their wellbeing and had contacted the Coastguard.

Search for missing person

Date: Friday 19 February 2021

Location: Falmouth Inner Harbour and Falmouth Bay

Weather: S 6, rough sea and moderate swell, good visibility, overcast

All-weather lifeboat: (launched 2.17am / Returned 3.34am) – Jonathon Blakeston (Coxswain), Neil Capper, Elliot Holman, Nick Head, Jamie Connolly, Andy Edwards, Lloyd Barron

Inshore lifeboat: (launched 2.14am / returned 3.36am) – James Wakefield (Helmsman), Tom Bird, Tamara Brookes, Joe Amps

Details:

At 2.03am Falmouth Coastguard requested that Falmouth all-weather and inshore lifeboats be launched to assist the police in multi-agency search for an elderly person who had not returned from an early evening walk in the Swanpool area. The Falmouth and Porthoustock Coastguard Cliff Rescue Teams, Cornwall Search & Rescue Team, Police Units, Police helicopter and members of the public were also involved in the search.

The inshore lifeboat launched from her slipway at 2.14am and commenced its search of the Falmouth Inner Harbour at 2.16am. The all-weather lifeboat left its pontoon berth at 2.17am, and headed out of the harbour commencing its search of from Pendennis Point to Pennance Point at 2.30am. The inshore lifeboat searched the Inner Harbour and the Penryn River up to Ponsharden before retracing its steps, completing its search at 3.30am. The all-weather lifeboat had completed its search of the coastline by 3.20am. With nothing being found both lifeboats were released to return to their station.

The all-weather lifeboat was back alongside its pontoon berth at 3.34am, with the inshore lifeboat being back on its slipway by 3.36am. Both lifeboats were refuelled and made ready for service by 3.55am.

An extensive search of the local area was carried out during the late evening and through the night. The missing person was eventually located at approx. 8.00am and was taken by Ambulance to Treliske Hospital for treatment. This was a massive multi-agency search supported by many people from the local community. Our thoughts are with the family and friends of the missing person at this difficult time.

Missing Child

Date: Sunday 16 February 2020

Location: between Castle Beach and Pendennis Headland

Weather: SW 5, moderate sea and swell, excellent visibility, clear sky

All-weather lifeboat: (launched 4.48pm / returned 4.52pm) – Luke Wills (Second Coxswain), Carl Beardmore, Dave Nicoll, Andy Jenkin, Jonathan Hackwell, Will Allen, Chris Simpson

Inshore lifeboat: (launched 4.44pm / Returned 4.59pm) – Sandy Procter (Helmsman), Tom Bird, Tamsin Mulcahy, Caden Harris

Details:

At 4.38pm Falmouth Coastguard requested that Falmouth inshore lifeboat be launched following a report of a missing child between Castle Beach and Pendennis Point in Falmouth. The Police and the Falmouth Coastguard Rescue Team were also tasked to assist with the search. The decision was made at 4.42pm to also launch the all-weather lifeboat to assist in the search.

The inshore lifeboat launched from her slipway at 4.44pm and arrived off Pendennis Point at 4.48pm to commence its search.

The all-weather lifeboat was manned by 4.48pm and was in the process of departing its pontoon berth at 4.50pm when the Coastguard advised both lifeboats that the child had been located safe and well on Castle Drive.

The all-weather lifeboat was secured back alongside its pontoon berth by 4.52pm. The inshore lifeboat was recovered onto its slipway by 4.5pm9, where it was refuelled and made ready for service by 5.10pm.

Persons cut off by the tide

Date: Sunday 23 February 2020

Location: between Pendennis Point and Swanpool Beach

Weather: W 3, slight sea and swell, good visibility, partly cloudy

Inshore lifeboat: (launched 5.35pm / returned 6.16pm) – Jamie Wakefield (Helmsman), Tamara Brookes, Tamsin Mulcahy, Lloyd Barron

Details:

At 5.28pm Falmouth Coastguard requested that Falmouth inshore lifeboat be launched after receiving a 999 call from two members of the public reporting that they were cut off by the tide somewhere between Pendennis Point and Swanpool Beach. The Falmouth Coastguard Rescue Team was also tasked to assist.

The inshore lifeboat launched from her slipway at 5.35pm and arrived off Pendennis Point at 5.40pm. The Coastguard Rescue Team positioned at Boscawen Fields quickly reported that they had located the casualties at the base of the cliff between Gyllyngvase and Swanpool Beaches. The inshore lifeboat arrived on scene at 5.44pm and it was decided that the best way was to recover the casualties was by the lifeboat. By 5.54pm both casualties were safely on board the lifeboat and they were then taken back to the lifeboat station where they were landed ashore at 6.10pm.

The inshore lifeboat was recovered onto its slipway at 6.16pm where the lifeboat was refuelled and made ready for service by 6.30pm.

The two casualties had been exploring the coast when they had become cut off by the tide. The initial information they had given was limited but thanks to the local knowledge of the Coastguard team and the lifeboat crew they had been quickly located and were recovered safely.

Fishing Vessel Toni Lou requiring assistance

Date: Thursday 5 March 2020

Location: 1 mile East of Portscatho

Weather: W 3, slight sea and no swell, excellent visibility, clear sky

All-weather lifeboat: (launched 7.17pm / returned 9.08pm) – Dave Nicoll (Deputy 2nd Coxswain), Andy Edwards, Tom Bird, Andy Jenkin, Sandy Procter, Neil Capper, Chris Simpson

Details:

At 7.06pm Falmouth Coastguard requested that Falmouth all-weather lifeboat be launched to assist the 9.8m Fishing Boat Toni Lou (FH6) which had two persons on board and had reported being adrift in Gerrans Bay.

The all-weather lifeboat left its pontoon berth at 7.17pm and having rounded Zone Point headed up to the East arriving on scene one-mile East of Portscatho at 7.42pm. A tow line was passed to the vessel and at 7.48pm the tow back to Falmouth commenced. On entering the entrance to Falmouth Harbour the tow was shortened and the vessel was strapped alongside the lifeboat so that it could be placed on the pontoons at Port Pendennis Marina. The fishing boat was safely secured alongside at the marina by 8.59pm and the lifeboat was released to return to its station.

The all-weather lifeboat was back alongside its pontoon berth by 9.08pm where the lifeboat was refuelled and made ready for service again by 9.25pm.

The fishing boat is used for scallop diving and had been on passage from Mylor to Fowey when it had suffered an issue with its fuel system. While it had been drifting out to sea and in no immediate danger, there were no other vessels in the area, so the decision was made to request the lifeboat to tow the vessel back to port before the weather deteriorated.

Kayaker in difficulty

Date: Thursday 12 March 2020

Location: off Porthcurnic Beach, Gerrans Bay

Weather: W 5-6, moderate sea and swell, good visibility, squally showers

Inshore lifeboat: (launched 1.32pm / returned 2.55pm) – Jamie Wakefield (Helmsman), Tom Bird, Caden Harris, Lloyd Barron

Details:

At 1.25pm Falmouth Coastguard requested that Falmouth inshore lifeboat be launched after they received several 999 calls reporting a Kayaker in difficulties off Porthcurnic Beach, near Portscatho. The Portscatho Coastguard Rescue Team was also tasked to assist.

The inshore lifeboat launched from her slipway at 1.32pm and having rounded Zone Point headed up to the east arriving on scene off Portscatho at 1.46pm. The casualty was spotted clinging to his kayak about a mile offshore and was quickly recovered to the inshore lifeboat, where they were found to be suffering from severe hypothermia. The decision was made to abandon the kayak and at 1.50pm the inshore lifeboat headed back towards Falmouth at best possible speed given the weather conditions. Arriving back at the lifeboat station at 2.06pm the inshore lifeboat was moored alongside the all-weather lifeboat where the casualty was placed on a stretcher and moved into the wheelhouse of the all-weather lifeboat to be assessed by ambulance paramedics, before being transferred to an ambulance and taken to Treliske Hospital.

The inshore lifeboat was recovered onto its slipway at 2.55pm where the lifeboat was refuelled and made ready for service by 3.20pm.

The kayaker had launched from Porthcurnic Beach and got into difficulty when they lost their paddle and then ended up in the water. They are believed to have been in the water for up to two hours. Without the reports from the public and the actions of the lifeboat crew the outcome could have had a tragic conclusion.

Person cut off by the tide

Date: Friday 3 April 2020

Location: west side of Pendennis Point

Weather: variable 1, calm sea and no swell, excellent visibility, clear skies

Inshore lifeboat: (launched 11.23pm / returned 0.01am) – Jamie Wakefield (Helmsman), Tom Bird, Joe Amps, Jemima Henstridge-Blows

Details:

At 11.10pm Falmouth Coastguard requested that Falmouth inshore lifeboat be launched following reports of a person cut off by the tide on the west side of Pendennis Point. The Falmouth Coastguard Rescue Team was also tasked to assist.

The inshore lifeboat launched from her slipway at 11.23pm and having round Pendennis Point, arrived on scene on the west side of the headland at 11.30pm. The casualty had already been located by the Coastguard team on a rock island which they were unable to reach. The inshore lifeboat approached the rock and recovered the casualty onto the lifeboat before they were taken to the east side of the point where they were landed on the beach near Crab Quay. Once the casualty was safely ashore at 11.50pm the inshore lifeboat was released to return to its station.

The inshore lifeboat was recovered onto its slipway at 0.01am where the lifeboat was refuelled and made ready for service by 0.15am.

The rock island is only accessible from the shore at low tide so with two hours to go to high water the casualty would have had to remain on the island until nearly daybreak. Why the casualty was on the rock is not known.

Persons cut off by the tide

Date: Sunday 10 May 2020

Location: Maenporth Beach

Weather: NNE 3, calm sea and no swell, good visibility, overcast with rain

Inshore lifeboat: (launched 5.12pm / returned 5.30pm) – Elliot Holman (Helmsman), Tom Bird, Tamara Brookes, Lloyd Barron

Details:

At 5.04pm Falmouth Coastguard requested that Falmouth inshore lifeboat be launched following reports of three people cut off by the tide at Maenporth Beach. The Falmouth and Porthoustock Coastguard Rescue Teams were also tasked to assist.

The inshore lifeboat launched from her slipway at 5.12pm and was about to round Pendennis Point when at 5.19pm the coastguard advised that the three casualties had made it ashore so the inshore lifeboat was released to return to its station while the Falmouth Coastguard Rescue Team continued to Maenporth.

The inshore lifeboat was recovered onto its slipway at 5.30pm where the lifeboat was refuelled and made ready for service by 5.45pm.

The three casualties had made their own way back to the beach, although a little wet, it was confirmed that they were safe and well. This was the second call out for the inshore lifeboat to people cut off by the tide at Maenporth during the bank holiday weekend, the previous incident on the Friday was cancelled when the casualty made it safely ashore just as the inshore lifeboat was about to be launched from her slipway.

Yacht in danger of drifting ashore

Date: Tuesday 19 May 2020

Location: Penarrow Point, Carrick Roads

Weather: W 2, calm sea and no swell, good visibility, overcast

Inshore lifeboat: (launched 3.44pm / returned 4.19pm) – Jamie Wakefield (Helmsman), Tom Bird, Tamara Brookes, Lloyd Barron

Details:

At 3.36pm Falmouth Coastguard requested that Falmouth inshore lifeboat be launched following a call from a 9m single handed yacht reporting it had suffered engine failure and was drifting onto Penarrow Point in the Carrick Roads. The Falmouth Coastguard Rescue Teams were also tasked to assist.

The inshore lifeboat launched from her slipway at 3.44pm and headed out of the Inner Harbour and up the Carrick Roads towards Penarrow Point. At 3.47pm the yacht was located underway between Trefusis Point and Penarrow Point, having managed to get its engine going, so it was agreed that the inshore lifeboat would escort the yacht to its berth in Falmouth Marina in the Penryn River. While passing Trefusis Point its engine failed again so the yacht was taken in tow by the inshore lifeboat. At 4.09pm as the boats entered the Penryn River the tow was handed over to the Marina Workboat, the inshore lifeboat being then released to return to its station. The Falmouth Coastguard Rescue Team which had initially gone to Mylor Harbour relocated to Falmouth Marina to meet the casualty.

The inshore lifeboat was recovered onto its slipway at 4.19pm where the lifeboat was refuelled and made ready for service by 4.30pm.

The yacht is believed to have been on passage from Mylor to Falmouth Marina when its engine had failed. Being close to the rocks of Penarrow Point with limited wind to sail by the skipper had requested assistance.

Motorboat broken down

Date: Wednesday 20 May 2020

Location: between Mylor and Falmouth Docks

Weather: S 2, calm sea and no swell, fair visibility, partly cloudy

Inshore lifeboat: (launched 9.07pm / returned 10.05pm) – Nick Head (Helmsman), Chris Simpson, Jack Williams, Jemima Henstridge-Blows

Details:

At 8.59pm Falmouth Coastguard requested that Falmouth inshore lifeboat be launched following a 999 call from a 6m motorboat reporting that it had broken down with four persons on board in the Carrick Roads somewhere between Mylor and Falmouth Docks. The Falmouth Coastguard Rescue Team was also tasked to assist.

The Inshore Lifeboat launched from her slipway at 9.07pm and headed up the Carrick Roads searching from Trefusis Point towards Mylor. There was no sign of the boat, so the inshore lifeboat was tasked to search the inner harbour but shortly after arriving back in the harbour the Coastguard confirmed they had managed to contact the first informant again and it appeared that the boat was further up the Carrick Roads near Restronguet Creek. The inshore lifeboat headed back up the Carrick Roads and at 9.29pm located the boat off the Carick Carlys Rock. The boat was taken in tow back to the Grove Place Dinghy Park where it was safely moored at 10.01pm, the inshore lifeboat being then released to return to its station.

The inshore lifeboat was recovered onto its slipway at 10.05pm where the lifeboat was refuelled and made ready for service by 10.40pm.

It appears that the boat had run out of fuel. It had no navigation lights or a VHF Radio, only one lifejacket between the four occupants and contained several empty alcohol containers. The Coastguard Rescue Team met the casualties on their arrival at the Dinghy Park and provided suitable safety advice.

'

Person's cut off by the tide

Date: Monday 25 May 2020

Location: near Shag Rock, Mawnan Shear

Weather: variable 1, calm sea and no swell, excellent visibility, clear sky

All-weather lifeboat: (launched 11.02pm / returned 11.59pm) – Carl Beardmore (Deputy 2nd Coxswain), Dave Nicoll, Jonathan Hackwell, Elliot Holman, Jamie Connolly, Will Allen, Joe Amps

Inshore lifeboat: (launched 10.59pm / Returned 11.49pm) – Luke Wills (Helmsman), Tom Bird, Jamie Wakefield, Tamara Brookes

Details:

At 1-.51pm Falmouth Coastguard requested that Falmouth all-weather and inshore lifeboats be launched following a 999 call from a family group who were cut off by the tide near Shag Rock at Mawnan Shear. The Falmouth Coastguard Rescue Team was also tasked to assist.

The inshore lifeboat launched from her slipway at 10.59pm followed shortly afterwards by the all-weather lifeboat which left its pontoon berth at 11.02pm. Having rounded Pendennis Point both lifeboats headed across Falmouth Bay, the inshore lifeboat arriving on scene at 11.10pm quickly locating the casualties on the shoreline.

The all-weather lifeboat arrived on scene off August Rock at 11.21pm just as the inshore lifeboat confirmed that all six casualties had been recovered. The all-weather lifeboat stood by in the mouth of the Helford River while the inshore lifeboat landed the casualties into the care of the Coastguard Team at Durgan Beach. Once the casualties were safely ashore at 11.33pm both Lifeboats were released to return to their station.

The inshore lifeboat was recovered onto its slipway at 11.49pm and the all-weather lifeboat was back alongside its pontoon berth at 11.59pm. Both lifeboats were refuelled and ready for service by 0.14am.

The three adults and three young children had been exploring the coast when they had become cut off by the tide. They had initially planned to wait for the tide to recede but became concerned for their children with the onset of darkness as the temperature so wisely called for assistance. They were very appreciative of the help they were given by the volunteer lifeboat crew and coastguards.

Fishing Boat Samuel J broken down

Date: Thursday 28 May 2020

Location: 4 miles South of Dodman Point

Weather: ESE 4-5, moderate sea and heavy swell, excellent visibility, clear sky

All-weather lifeboat: (launched 3.22pm / returned 6.10pm) – Jonathon Blakeston (Coxswain), Carl Beardmore, Dave Nicoll, Will Allen, Andy Edwards, Wendy Rabett, Lloyd Barron

Boarding boat: (launched 5.30pm / Returned 6.05pm) – Luke Wills (Helmsman), Andy Jenkin

Details:

At 3.09pm Falmouth Coastguard requested that Falmouth all-weather lifeboat be launched to assist the 6.5m Fishing Boat Samuel J with one person on board which had broken down 4 miles South of Dodman Point.

The all-weather lifeboat left its pontoon berth at 3.22pm and having rounded Zone Point, set a course to the east arriving on scene at 3.55pm. Due to heavy swell that was running a drogue was passed to the casualty before a tow line was connected, the tow to Falmouth commencing at 4.15pm.

The lifeboat and casualty passed St Anthony Lighthouse at 5.15pm and once off the Docks the towline was recovered and the fishing boat was strapped alongside the lifeboat so it could be placed on a mooring in the Inner Harbour. The stations boarding boat was launched at 5.30pm to assist with mooring the boat. The fishing boat was safely placed on the visitors mooring in the harbour by 6.00pm and the all-weather lifeboat and the boarding boat were released to return to their station.

The boarding boat was back on its pontoon dock by 6.05pm with the all-weather lifeboat back alongside its pontoon berth by 6.10pm where it was refuelled and made ready for service by 6.25pm.

The newly purchased fishing boat was on passage from Brixham to Malpas in the Truro River when it broke down. Due to the heavy swell the skipper had been unable to rectify the fault so had asked for assistance.

Person needing medical evacuation

Date: Sunday 14 June 2020

Location: Pennance Point

Weather: SW 4, slight sea and no swell, good visibility, partly cloudy

Inshore lifeboat: (launched 3.38pm / returned 4.03pm) – Jamie Connolly (Helmsman), Tom Bird, Andy Edwards, Lloyd Barron

Details:

At 3.30pm Falmouth Coastguard requested that Falmouth inshore lifeboat be launched to assist the Falmouth Coastguard Rescue Team and ambulance paramedics with the evacuation of a casualty from Pennance Point.

The inshore lifeboat launched from her slipway at 3.38pm and having rounded Pendennis Point headed across Falmouth Bay arriving on scene at 3.48pm. Soon after arriving it was confirmed that the casualty had been transferred to Swanpool Beach by another vessel so at 3.50pm the inshore lifeboat was released to return to its station.

The inshore lifeboat was recovered onto its slipway at 4.03pm where the lifeboat was refuelled and made ready for service by 4.20pm.

The casualty is believed to have been taking part in coasteering when they had injured themselves. Due to the considerable time the casualty had been in the water the paramedics were concerned that they were becoming hypothermic and given the remote location extraction by boat was the fastest option.

Dinghy with rudder failure on a lee shore

Date: Sunday 21 June 2020

Location: Carrick Roads between St Mawes and St Just

Weather: W 4-5, slight sea and no swell, excellent visibility, partly cloudy

Inshore lifeboat: (launched 6.34pm / returned 7.29pm) – Elliot Holman (Helmsman), Tom Bird, Tamara Brookes, Chris Simpson

Details:

At 6.26pm Falmouth Coastguard requested that Falmouth inshore lifeboat be launched following a 999 call from the occupants of a 4.3m sailing dinghy which had suffered rudder failure and was being blown towards the shoreline in the Carrick Roads between St Mawes and St Just.

The inshore lifeboat launched from her slipway at 6.34pm and headed up the Carrick Roads arriving on scene at 6.40pm. The occupants of the dinghy had managed to secure themselves to one of the ski markers so a tow line was passed to the dinghy and it was taken back to the Penryn River where it was safely place on the slipway at Trevissome at 7.16pm, the inshore lifeboat being released to return to its station.

The inshore lifeboat was recovered onto its slipway at 7.29pm where the lifeboat was refuelled and made ready for service by 7.45pm.

The rudder had failed on the dinghy and they had been unable to make headway against the moderate Westerly winds as they did not have an auxiliary outboard. Suitable safety advice was given to the casualties in respect to having lifejackets and a handheld radio.

Report of boat adrift

Date: Thursday 25 June 2020

Location: off Porthoustock

Weather: ESE 1, slight sea and no swell, good visibility, partly cloudy with occasional rain

Inshore lifeboat: (launched 10.41am / returned 12.10pm) – Neil Capper (Helmsman), Tamara Brookes, Caden Harris, Lloyd Barron

Details:

At 10.33am Falmouth Coastguard requested that Falmouth inshore lifeboat be launched following reports of a boat adrift off Porthoustock. The Porthoustock Coastguard Cliff Rescue Team was also tasked to assist.

The inshore lifeboat launched from her slipway at 10.41am and having rounded Pendennis Point headed across Falmouth Bay to the initial position given off Porthoustock. A search commenced to locate the boat which was found at 11.05am just under a mile to the east of Porthoustock under tow by another vessel. The drifting boat was found to be a 3.5m Rigid Inflatable, its outboard was raised but its kill cord was still attached and there was a lifejacket on board. The inshore lifeboat took over the tow taking the vessel to Porthoustock where it was handed into the care of the coastguard team

The inshore lifeboat was then tasked at 11.35am to carry out a shoreline search from the Manacles to Nare Point looking for a potential person in the water. At 11.54am the coastguard confirmed that the boat had gone missing from the Helford, so the inshore lifeboat was released to return to its station.

The inshore lifeboat was recovered onto its slipway at 12.10pm where the lifeboat was refuelled and made ready for service by 12.30pm.

It turned out that the owner of the boat had been searching the Helford River for his boat unaware that it had in fact drifted out into Falmouth Bay and down the coast towards the Manacles. The assistance of the passing boat which had initially taken the boat in tow was very much appreciated.

Swimmers requiring assistance

Date: Sunday 28 June 2020

Location: Porthbeor Beach

Weather: WSW 5-6, moderate sea and slight swell, good visibility, partly cloudy

Inshore lifeboat: (launched 12.19pm / returned 1.15pm) – Nick Head (Helmsman), Adam West, Tamara Brookes, Joe Amps

Details:

At 12.03pm Falmouth Coastguard requested that Falmouth inshore lifeboat be launched following reports of two swimmers requiring assistance at Porthbeor Beach. The Portscatho Coastguard Cliff Rescue Team was also tasked to assist.

The inshore lifeboat launched from her slipway at 12.19pm and having rounded Zone Point headed east arriving on scene at 12.31pm. One swimmer was located in a cove at the eastern end of the beach, being assisted by a member of the public on a paddle board. The casualty was then transferred to the inshore lifeboat by the paddle boarder. It was then confirmed that the second swimmer had landed further along the beach and had made their own way from the beach up to the cliff path above. The swimmer on the inshore lifeboat was checked over by a casualty care trained crewmember and while they were cold, they were in no immediate danger. Once it was confirmed that no other swimmers were in difficulties the swimmer was taken on to St Mawes where they were landed at 1.01pm, the inshore lifeboat then being released to return to its station.

The inshore lifeboat was recovered onto its slipway at 1.15pm where the lifeboat was refuelled and made ready for service by 1.55pm.

In total there had been six triathlon swimmers who had departed from Towan Beach and were swimming along the coast to St Anthony Lighthouse and then back up the Percuil River. The six swimmers who were in three pairs had been acting on their own and were not part of an organised event when one swimmer had become exhausted in the prevailing weather conditions. Members of the public including two trained volunteer lifeguards had become concerned for their wellbeing and had therefore taken a paddle board down to Porthbeor Beach before heading over to the cove to assist the swimmer. Initial first aid was given by the experienced volunteer lifeguard before the arrival of the Inshore Lifeboat. This incident could have been much worse if it had not been for the intervention of members of the public.

Fishing Boat Dawn Star with engine failure

Date: Tuesday 30 June 2020

Location: 14 miles ESE of Lizard Point

Weather: WSW 5-6, moderate sea and slight swell, good visibility, partly cloudy

All-weather lifeboat: (launched 2.30pm / returned 3.30pm) – Jonathon Blakeston (Coxswain), Luke Wills, Dave Nicoll. Will Allen, Tamara Brookes, Andy Edwards, Jack Williams

Details:

At 2.21pm Falmouth Coastguard requested that Falmouth all-weather lifeboat be launched to assist the 10m fishing boat *Dawn Star* which had reported engine failure 14 miles ESE of Lizard Point.

The all-weather lifeboat left its pontoon berth at 2.30pm and headed out of the harbour. At 2.40pm as the lifeboat was passing Pendennis Point the Coastguard advised that the *Dawn Star* had managed to restart her engine but requested that the lifeboat continue until the vessel confirmed that their assistance was no longer required. At 3.05pm when the lifeboat was 4 miles South of Pendennis Point it was confirmed that the fishing boat was happy to return under its own power, so the lifeboat was released to return to its station.

The all-weather lifeboat was back alongside its pontoon berth by 3.30pm where it was refuelled and made ready for service by 3.40pm.

The Falmouth lifeboat was launched to assist this casualty as the Lizard lifeboat was already afloat searching 25 miles South of Lizard Point for an upturned boat which had been reported by a passing vessel.

People in the water from upturned boat

Date: Thursday 2 July 2020

Location: off Mylor Harbour

Weather: W 4, slight sea and no swell, good visibility, cloudy

Inshore lifeboat: (launched 1.22pm / returned 1.30pm) – Jamie Wakefield (Helmsman), Tom Bird, Jack Williams, Lloyd Barron

Details:

At 1.15pm Falmouth Coastguard requested that Falmouth inshore lifeboat be launched following a Mayday call reporting four people in the water from an upturned boat off Mylor Harbour. The Falmouth Coastguard Cliff Rescue Team was also tasked to assist.

The inshore lifeboat launched from her slipway at 1.22pm and was off the Queens Wharf when at 1.25pm the Coastguard advised that the people had been recovered from the water, so the inshore lifeboat was released to return to its station.

The inshore lifeboat was recovered onto its slipway at 1.30pm where the lifeboat was refuelled and made ready for service by 1.50pm.

Nearby vessels were able to recover the casualties from the water and assist with the recovery of the boat. As no medical assistance was required the inshore lifeboat was released while the Coastguard Mobile continued to Mylor to obtain additional information

Two paddle boarders being blown offshore

Date: Sunday 5 July 2020

Location: Swanpool Beach

Weather: W 5-6, slight sea and no swell, excellent visibility, partly cloudy

Inshore lifeboat: (launched 2.14pm / returned 2.20pm) – Luke Wills (Helmsman), Jamie Connolly, Tamara Brookes, Jack Williams

Details:

At 2.07pm Falmouth Coastguard requested that Falmouth inshore lifeboat be launched following reports of two paddle boarders being blown offshore at Swanpool Beach. The Falmouth Coastguard Cliff Rescue Team was also tasked to assist.

The inshore lifeboat launched from her slipway at 2.14pm and was just starting to head out of the harbour when at 2.16pm the Coastguard advised that the paddle boarders were being assisted ashore and no further assistance was required so the inshore lifeboat was released to return to its station.

The inshore lifeboat was recovered onto its slipway at 2.20pm where the lifeboat was refuelled and made ready for service by 2.35pm.

The two paddle boarders were being blown offshore in the strong westerly winds. They were assisted ashore by members of the local water sports centre. The Coastguard Team continued to Swanpool Beach to speak to the casualties and provide safety advice.

Sailing Dinghy being blown onto rocks

Date: Friday 10 July 2020

Location: between St Anthony Lighthouse and Little Molunan

Weather: NW 4, slight sea and swell, excellent visibility, partly cloudy

Inshore lifeboat: (launched 12.35pm / returned 1.39pm) – Jamie Wakefield (Helmsman), Tamara Brookes, Andy Edwards, Lloyd Barron

Details:

At 12.28pm Falmouth Coastguard requested that Falmouth inshore lifeboat be launched to assist a 4.8m sailing dinghy with three people and a dog on board which had suffered engine failure and was in danger of being blown onto the rocks between Little Molunan Beach and Shag Rock below St Anthony Lighthouse.

The inshore lifeboat launched from her slipway at 12.35pm and having rounded the docks was advised by the Coastguard that the boat was now on the rocks. The inshore lifeboat arrived on scene at 12.42pm, locating the boat in one of the rock gullies near Shag Rock. A lifeboat crew member was placed ashore to assist in moving the vessel to a location where it could be towed clear of the shore. Once refloated the boat and its occupants were towed to St Mawes Harbour where at 1.26pm the tow was handed over to a boat from the local sailing club, the inshore lifeboat then being released to return to its station.

The inshore lifeboat was recovered onto its slipway at 1.39pm where the lifeboat was refuelled and made ready for service by 2.00pm.

The sailing dinghy was close to the shore between St Anthony Lighthouse and Little Molunan Beach when its outboard had failed and was therefore unable to sail clear so ended up on the rocks.

Report of person on upturned vessel

Date: Tuesday 14 July 2020

Location: off the mouth of the Helford

Weather: NW 5, slight sea and swell, good, overcast

Inshore lifeboat: (launched 2.22pm / returned 13.45pm) – Jamie Wakefield (Helmsman), Tom Bird, Tamara Brookes, Lloyd Barron

Details:

At 2.16pm Falmouth Coastguard requested that Falmouth inshore lifeboat be launched following a report from the Nare Head NCI Lookout reporting a person on an upturned vessel at the mouth of the Helford. The Porthoustock Coastguard Cliff Rescue Team was also tasked to assist.

The inshore lifeboat launched from her slipway at 2.22pm and having rounded Pendennis Point headed across Falmouth Bay, arriving off the mouth of the Helford River at 2.33pm. One adult and four children were located on the shoreline with a kayak while a second adult was located on a nearby rigid inflatable, along with an open canoe. Two volunteer lifeboat crew members were dropped ashore to check on the casualties on the shoreline while the other casualty was taken on board the inshore lifeboat along with their canoe. The four children were then transferred from the shore to the inshore lifeboat using the open canoe. The children remained on the inshore lifeboat while it towed the open canoe back into Gillan, while the adults followed on in their kayak. Once they were all safely ashore in Gillan Creek at 3.24pm the inshore lifeboat was then released to return to its station.

The inshore lifeboat was recovered onto its slipway at 3.45pm where the lifeboat was refuelled and made ready for service by 4.15pm.

The family members had been exploring the waters off Gillan when their Canadian style open canoe had capsized throwing one adult and four of the children into the water. The adult on the kayak helped the children ashore while the other adult remained with the upturned canoe and was assisted by a passing rigid inflatable.

Dive Boat broken down

Date: Wednesday 15 July 2020

Location: off Nare Head, Gerrans Bay

Weather: NW 4, slight sea and swell, good visibility, partly cloudy

All-weather lifeboat: (tasked 1.07pm / stood down 1.55pm) – Jonathon Blakeston (Coxswain), Luke Wills, Dave Nicoll, Tom Bird, Jamie Wakefield, Will Allen, James Murray

Details:

At 1.07pm Falmouth Coastguard requested that Falmouth all-weather lifeboat be brought to immediate readiness following a call from a 10m diver boat which had reported breaking down off Nare Head, Gerrans Bay.

The all-weather lifeboat remained alongside its pontoon with the engines running while it awaited an update from the Coastguard. The dive boat was at anchor and all its divers were on board the vessel so there was no immediate risk. A rigid inflatable proceeded from Mylor Harbour to assist the vessel. At 1.53pm the skipper of the dive boat confirmed that it was under tow and that they were happy with the progress so it was agreed that the lifeboat could be stood down.

The all-weather lifeboat's engines were shut down and it was made ready for service by 1.55pm.

Occupants of a broken down boat trying to swim ashore

Date: Wednesday 15 July 2020

Location: off Ski Beach, Carrick Roads

Weather: NW 3-4, slight sea and no swell, good, partly cloudy

Inshore lifeboat: (launched 6.57pm / re-tasked 7.36pm) – Jamie Wakefield (Helmsman), Tom Bird, Tamara Brookes, Chris Simpson

Details:

At 6.49pm Falmouth Coastguard requested that Falmouth inshore lifeboat be launched to assist the occupants of a 4m open boat broken down off Ski Beach who had entered the water to try and get ashore.

The inshore lifeboat launched from her slipway at 6.57pm and headed up the Carrick Roads. While on route the inshore lifeboat was advised that the three occupants had made it to the shore. The inshore lifeboat arrived on scene at 7.03pm and having checked that everyone was uninjured, the boat and its occupants were towed clear of the beach at 7.14pm and were taken back to Mylor Harbour where they safely landed ashore at 7.33pm, the inshore lifeboat being released to return to its station.

As the inshore lifeboat was heading back out into the Carrick Roads they became aware that the sailing dinghy that had been accompanying the first casualty was not making any headway so at 7.36pm they diverted to assist the sailing dinghy.

Sailing Dinghy struggling to make headway

Date: Wednesday 15 July 2020

Location: St Just Pool, Carrick Roads

Weather: NW 3-4, slight sea and no swell, good, partly cloudy

Inshore lifeboat: (tasked 7.36pm / re-tasked 8.07pm) – Jamie Wakefield (Helmsman), Tom Bird, Tamara Brookes, Chris Simpson

Details:

Falmouth inshore lifeboat had just towed an open boat and its occupants into Mylor Harbour and was heading back out into the Carrick Roads at 7.36pm when they became aware that the sailing dinghy which had been accompanying the earlier casualty was struggling to make headway against the moderate North westerly wind.

The inshore lifeboat arrived on scene at the St Just Pool at 7.39pm and found that the dinghy was having rigging issues, so it was taken in tow. The dinghy and its two occupants were towed to Mylor Harbour where it was safely secured at 7.57pm, the inshore lifeboat once again being released to return to its station.

As the inshore lifeboat was approaching Falmouth Harbour at 8.07pm they became aware of a 3m rigid inflatable broken down between the Eastern Arm of the Docks and Trefusis Point and proceeded to assist.

Small rigid inflatable broken down

Date: Wednesday 15 July 2020

Location: between Eastern Arm and Trefusis Point

Weather: NW 3-4, slight sea and no swell, good, partly cloudy

Inshore lifeboat: (tasked 8.07pm / returned 8.20pm) – Jamie Wakefield (Helmsman), Tom Bird, Tamara Brookes, Chris Simpson

Details:

Falmouth inshore lifeboat was returning to station from a previous incident when at 8.07pm they became aware of a 3m rigid inflatable with one occupant which had clearly broken down, so they proceeded to assist.

The inshore lifeboat arrived on scene at 8.09pm and took the boat in tow. It was towed into the Inner Harbour and once off the Port Pendennis Marina the boat was released at 8.17pm so the occupant could row to its berth, the inshore lifeboat once again being released to return to its station.

The inshore lifeboat was recovered onto its slipway at 8.20pm where the lifeboat was refuelled and made ready for service by 8.40pm.

The inshore lifeboat had in just over an hour assisted three boats and their six occupants. Suitable safety advice was given where applicable.

Report of collision with people in the water

Date: Friday 17 July 2020

Location: off Loe Beach, Carrick Roads

Weather: W 2, slight sea and no swell, excellent visibility, clear sky

Inshore lifeboat: (launched 10.31pm / returned 11.22pm) – Neil Capper (Helmsman), Tom Bird, Nick Head, Jemima Henstridge-Blows

Details:

At 10.22pm Falmouth Coastguard requested that Falmouth inshore lifeboat be launched following a collision between boats and reports of a person in the water. The Falmouth Coastguard Cliff Rescue Team and the Police were already on scene.

The inshore lifeboat launched from her slipway at 10.31pm and headed up the Carrick Roads arriving on scene at Loe Beach at 10.40pm. The Coastguard on scene confirmed that the person who had been in the water had already been assisted ashore so the inshore lifeboat stood by while two other occupants made it ashore in a tender. A check of the local area was made to ensure there were no other casualties. Once this was complete the inshore lifeboat crew went on board the motorboat to check its condition and to make sure it was secure. Once this was complete at 11.11pm the inshore lifeboat was then released to return to its station.

The inshore lifeboat was recovered onto its slipway at 11.22pm where the lifeboat was refuelled and made ready for service by 11.45pm.

The motorboat which had three people on board had been in collision with several other moored boats off Loe Beach. Several of the occupants were under the influence of alcohol and one had then ended up in the water following an apparent altercation. Suitable safety advice was given by the Coastguard Team and the Police.

Cries for help

Date: Sunday 19 July 2020

Location: Penryn River between Greenbank and Little Falmouth

Weather: SW 2, calm sea and no swell, fair visibility, overcast with drizzle

All-weather lifeboat: (launched 1.14am / returned 2.12am) – Carl Beardmore (Deputy 2nd Coxswain), Dave Nicoll, Sandy Procter, Tamsin Mulcahy, Andy Edwards, Caden Harris, Jemima Henstridge-Blows

Inshore lifeboat: (launched 1.03am / returned 2.30am) – Neil Capper (Helmsman), Nick Head, Tamara Brookes, Joe Amps

Details:

At 0.55am Falmouth Coastguard requested that Falmouth inshore lifeboat be launched following reports of cries for help coming from the Penryn River between Greenbank and Little Falmouth. The Falmouth Coastguard Cliff Rescue Team was also tasked to assist.

The inshore lifeboat launched from her slipway at 1.03am and headed up Falmouth Inner Harbour and into Penryn River. At 1.14am the inshore lifeboat located the casualty on the island pontoon off Little Falmouth. The casualty had been recovered from the water by a member of the public and two of the volunteer lifeboat crew members were placed onto the pontoon to provide casualty care.

At 1.14am the all-weather lifeboat was requested to launch to provide additional support and left its pontoon berth at 1.20am with a doctor on board who is one of the lifeboat volunteer crew. The all-weather lifeboat arrived on scene at 1.30am and moored alongside the pontoon to provide additional medical support. The casualty was transferred to the all-weather lifeboat and at 1.40am both lifeboats headed back down the Penryn River towards Prince of Wales Pier where an ambulance was waiting. By 2.06am the casualty had been landed ashore into the care of the Ambulance Paramedics and the all-weather lifeboat was released to return to its station. The inshore lifeboat was requested to return to the original scene to check on the casualty's vessel and at 2.21am having ensured that the yacht and its tender were secure the inshore lifeboat was also released to return to station.

The all-weather lifeboat was back alongside its pontoon berth at 2.12am, with the inshore lifeboat being recovered onto its slipway at 2.30am, with both lifeboats being refuelled and made ready for service by 3.00am.

The casualty had been returning to their moored yacht in its tender when it appears, they fell into the water. Luckily several people heard the casualties cries for help and one member of the public had the presence of mind to launch their paddle board and cross the river where they were able to drag the casualty onto the pontoon island. Without their actions the outcome of this incident could have been significantly different, helping to save this casualties life.

Yacht caught on fishing gear

Date: Sunday 19 July 2020

Location: 0.5 miles East of Nare Head, Helford

Weather: N 3-4, slight sea and no swell, excellent visibility, partly cloudy

Inshore lifeboat: (launched 6.54am / returned 10.14am) – Neil Capper (Helmsman), Tamara Brookes, Tamsin Mulcahy, Caden Harris

Details:

At 6.42am Falmouth Coastguard requested that Falmouth inshore lifeboat be launched following a call from a 13m yacht reporting that they were caught on fishing gear near Nare Head at entrance to the Helford.

The inshore lifeboat launched from her slipway at 6.54am and having rounded Pendennis Point headed across Falmouth Bay arriving on scene at 7.10am. Lifeboat crew from the inshore lifeboat were put on board the yacht to try and assist in freeing it from the fishing gear. The yacht had deployed its anchor, and this combined with the strong ebb tide made the task more difficult. By 8.40am the yacht was freed from the fishing gear, but the remaining rope meant that the yachts engine could not be used, and its rudder was affected, so the yacht was taken in tow. The yacht was taken to Mylor Harbour where it was safely handed over at 10.02am. The inshore lifeboat was then released to return to its station.

The inshore lifeboat was recovered onto its slipway at 10.14am where the lifeboat was refuelled and made ready for service by 10.45am.

The skipper of the yacht which had two people, a dog and cat on board had wisely requested assistance when they had been unable to free themselves from the fishing gear.

Motorboat with Engine Failure

Date: Sunday 19 July 2020

Location: 0.3 miles South of Porthmellin Head

Weather: N 3-4, slight sea and no swell, excellent visibility, partly cloudy

Inshore lifeboat: (launched 11.12am / returned 12.04pm) – Neil Capper (Helmsman), Tamara Brookes, Tamsin Mulcahy, Chris Simpson

Details:

At 11.05am Falmouth Coastguard requested that Falmouth inshore lifeboat be launched following a DSC distress call from a 6m Motorboat which was believed to have suffered engine failure.

The inshore lifeboat launched from her slipway at 11.12am and having rounded Zone Point headed up to the east towards Porthmellin Head. At 11.24am the boat with two persons on board was located underway heading back towards Zone Point. The boat was escorted back to St Mawes Harbour where at 11.55am the inshore lifeboat was released to return to its station.

The inshore lifeboat was recovered onto its slipway at 12.04pm where the lifeboat was refuelled and made ready for service by 12.35pm.

The Coastguard had issued a Mayday relay broadcast but with no vessels in the immediate vicinity and no direct contact with the vessel it was decided to launch the inshore lifeboat. The exact nature of the distress was not established but all on board the motorboat appeared fine. This was the third service for the inshore lifeboat in less than 12 hours.

Motorboat Ocean Drive taking on water

Date: Tuesday 21 July 2020

Location: ½ mile East of Killigerran

Weather: variable 1, calm sea and no swell, excellent visibility, partly cloudy

All-weather lifeboat: (launched 11.28am / returned 12.43pm) – Jonathon Blakeston (Coxswain), Luke Wills, Tom Bird, Sandy Procter, Will Allen, Tamara Brookes, Lloyd Barron

Details:

At 11.15am Falmouth Coastguard requested that Falmouth all-weather lifeboat be launched to assist the 10m Motorboat Ocean Drive which had reported flooding in its engine room. The vessel had two people on board and was ½ mile east of Killigerran Head.

The all-weather lifeboat left its pontoon berth at 11.28am and having rounded Zone Point headed up to the east. The lifeboat rendezvoused with the vessel which was making its way back towards Falmouth at 11.44am and two lifeboat crewmembers were transferred to the vessel. The source of the ingress had been identified and addressed so the lifeboat escorted the vessel into Falmouth Inner Harbour where it was moored on a visitors mooring. Thorough checks were carried out to ensure there were not any further sources of water ingress but with none evident the lifeboat was released at 12.34pm to return to its station.

The all-weather lifeboat was back alongside its pontoon berth by 12.43pm where it was refuelled and made ready for service by 12.50pm.

The water ingress had been caused due to a problem with a coolant pipe on one of its engines. This engine was shut down which eliminated the ingress and the water level in the engine compartment was reduced using the vessels pumps.

Person and Dog reported to be stuck in a gully

Date: Saturday 25 July 2020

Location: Trefusis Point

Weather: WSW 3, slight sea and no swell, good visibility, partly cloudy

Inshore lifeboat: (launched 3.05pm/ returned 3.35pm) – Jamie Connolly (Helmsman), Tamara Brookes, Chris Simpson, Cadan Harris

Details:

At 2.56pm Falmouth Coastguard requested that Falmouth inshore lifeboat be launched following reports that a person had become stuck in a gully while trying to recover their dog at Trefusis Point. The Falmouth Coastguard Cliff Rescue Team was also tasked to assist.

The inshore lifeboat launched from her slipway at 3.05pm and arrived on scene off Trefusis Point at 3.11pm. The casualty and their dog were quickly located and it was soon confirmed that they make their own way along the rocks and then back up to the coastal path. Once they were safely back on the cliff path, they were met by the Coastguard so at 3.24pm inshore lifeboat was released to return to its station.

The inshore lifeboat was recovered onto its slipway at 3.35pm where the lifeboat was refuelled and made ready for service by 3.55pm.

The first informant, who had been accompanying the casualty when he had gone to assist their dog, had been concerned that the casualty would not be able to get back out of the gully. Luckily the man and his dog had managed to find a route back onto rocks and were then able to re-join the coast path a little further along the point.

Report of person in water from inflatable

Date: Saturday 1 August 2020

Location: Pendennis Point

Weather: W 2, slight sea and no swell, good visibility, partly cloudy

Inshore lifeboat: (launched 8.47pm/ returned 10.03pm) – Elliot Holman, (Helmsman), Tom Bird, Lloyd Barron

Details:

At 8.40pm Falmouth Coastguard requested that Falmouth inshore lifeboat be launch following reports of an inflatable close to the rocks at Pendennis Point with a person in the water. The Falmouth Coastguard Cliff Rescue Team was also tasked to assist.

The inshore lifeboat launched from her slipway at 8.47pm and arrived on scene off Pendennis Point at 8.51pm. The small inflatable with two persons on board was located underway off Pendennis Point heading back towards the docks. Having had an initial discussion with the casualties the decision was taken to escort the boat back into the Inner Harbour. The Coastguard Cliff Rescue Team which had been stood down was retasked to meet the casualties on their arrival ashore. The dinghy was escorted back to the private slipway at Packet Quays where the casualties were met by the Coastguard team. At 9.56pm the inshore lifeboat was released to return to its station.

The inshore lifeboat was recovered onto its slipway at 10.03pm where the lifeboat was refuelled and made ready for service by 10.30pm.

Both the lifeboat crew and the coastguard team gave the casualties safety advice including highlighting the dangers of alcohol when boating.

Reports of Red Flare

Date: Sunday 2 August 2020

Location: Truro and Tresillian Rivers

Weather: WNW 2, calm sea and no swell, good visibility, partly cloudy

All-weather lifeboat: (launched 8.38pm / returned 10.57pm) – Dave Nicoll (Deputy 2nd Coxswain), Jonathan Hackwell, Neil Capper, Nick Head, Will Allen, Tamara Brookes, Andy Edwards

Inshore lifeboat: (launched 8.30pm / returned 10.45pm) – Elliot Holman (Helmsman), Tom Bird, Jamie Wakefield, Lloyd Barron

Details:

At 8.24pm Falmouth Coastguard requested that Falmouth inshore lifeboat be launched following reports of a red flare sighted in the area of Truro River. At 8.28pm it was decided to also launch the all-weather lifeboat to provide additional support including the use of its Y-boat inflatable. The Falmouth Coastguard Cliff Rescue Team was also tasked to assist.

The inshore lifeboat launched from her slipway at 8.30pm, with the all-weather lifeboat leaving its pontoon berth at 8.38pm. Both lifeboats headed up the Carrick Roads with the inshore lifeboat commencing its search of the Truro River from Malpas at 8.57pm while the all-weather lifeboat commenced its search of the River Fal from Turnaware Point at 8.58pm.

The inshore lifeboat completed its initial search of the Truro River from Malpas to Boscawen Park then headed back down the river to Malpas to meet the all-weather lifeboat's Y-boat inflatable which had been launched at Tolverne. Both the inshore lifeboat and the Y-boat then searched the Tresillian River up to St Clement's where the search had to be curtailed due to the ebbing tide. The all-weather lifeboat continued its search up the lower section of the Truro River working its way up to Woodbury Point before heading back down the river. With nothing being found at 10:05 the search was terminated. The Y-boat was successfully recovered at 10.15pm and both lifeboats headed back to station.

The inshore lifeboat was recovered onto its slipway at 10.45pm, with the all-weather lifeboat also being back alongside its pontoon berth by 10.57pm, both lifeboats being refuelled and made ready for service by 11.15pm.

The Coastguard had received numerous reports of the red flare. Initial indications were that it had been fired from the upper reaches of the Truro River, but further information obtained by the Falmouth Coastguard Cliff Rescue Team indicated that the flare was more likely to have come from the Tresillian River.

Search for missing person

Date: Monday 3 August 2020

Location: between Pennance Point and Rosemullion Head

Weather: WNW 2, calm sea and no swell, excellent visibility, clear sky

All-weather lifeboat: (launched 3.52am / Returned 6.40am) – Dave Nicoll (Deputy 2nd Coxswain), Jonathan Hackwell, Neil Capper, Nick Head, Will Allen, Tamara Brookes, Andy Edwards

Inshore lifeboat: (launched 3.50am / returned 5.50am) – Elliot Holman (Helmsman), Tom Bird, Jamie Wakefield, Lloyd Barron

Details:

At 3.40am Falmouth Coastguard requested that Falmouth's all-weather and inshore lifeboats be launched to assist in a multi-agency search for a missing person. The Falmouth, Porthoustock and Porthleven Coastguard Cliff Rescue Teams were also tasked to assist, along with the Police, Fire Brigade and Cornwall Search and Rescue Team.

The inshore lifeboat launched from her slipway at 3.50am, with the all-weather lifeboat leaving its pontoon berth at 3.52am. Having rounded Pendennis Point both lifeboats headed across the bay arriving off Pennance Point at 4.05am. A search of the shoreline was commenced from Pennance Point to Rosemullion Head using search lights and night sight equipment. By 5.30am both lifeboats had completed their search and with nothing being found the Coastguard released the lifeboats to return to their station.

The inshore lifeboat was recovered onto its slipway at 5.50am and was refuelled and made ready for service by 6.20am. The all-weather lifeboat had to standby in the Falmouth Inner Harbour while HMS Scott was berthed on the County Wharf but was back alongside its pontoon berth by 6.40am where it was refuelled and made ready for service by 7.00am.

The initial search had commenced shortly after midnight and had also involved the use of the Coastguard helicopter and its thermal imaging equipment. Both lifeboats were requested to launch when the search was extended to include the coastline on either side of Maenporth.

Concerns for person who had been in water

Date: Wednesday 5 August 2020

Location: St Mawes Harbour

Weather: SSW 4, slight sea and no swell, poor visibility, overcast with rain

Inshore lifeboat: (launched 2.15am/ returned 2.50am) – Jamie Connolly, (Helmsman), Tom Bird, Tamara Brookes, Joe Amps

Details:

At 2.07am Falmouth Coastguard requested that Falmouth inshore lifeboat be launched following a call from a person who had fallen into the water and had then got on board a moored vessel.

The inshore lifeboat launched from her slipway at 2.15am and arrived on scene in St Mawes Harbour at 2.23am. The casualty was located on board a small open boat. Having checked they did not need any medical assistance the casualty was returned to their yacht. At 2.36am the inshore lifeboat was released to return to its station.

The inshore lifeboat was recovered onto its slipway at 2.50am where the lifeboat was refuelled and made ready for service by 3.10am.

The casualty had been returning to his vessel when he had capsized his tender. Having secured the tender ashore, the casualty decided to swim back out to his vessel, but after swimming a short distance he decided it was too far and climbed on board an unoccupied vessel. He had planned to stay there overnight but decided to contact the coastguard when he started to succumb to the cold.

Yacht requiring assistance

Date: Wednesday 5 August 2020

Location: off Castle Beach, Falmouth Bay

Weather: SW 5, moderate sea and slight swell, fair visibility, overcast

All-weather lifeboat: (launched 4.31pm / returned 5.40pm) – Jonathon Blakeston (Coxswain), Tom Bird, Jamie Wakefield, Will Allen, Andy Edwards, Joe Amps, James Murray

Details:

At 4.20pm Falmouth Coastguard requested that Falmouth all-weather lifeboat be launched to assist a foreign yacht anchored off Castle Beach in Falmouth Bay. The yacht was having difficulties raising its anchor due to mechanical issues with its windless and there were concerns about the welfare of the two people on board due to the poor weather forecast.

The all-weather lifeboat left its pontoon berth at 4.31pm and having rounded Pendennis Point arrived on scene off Castle Beach at 4.42pm. Due to communication issues with the skipper a lifeboat crew member was placed on board the yacht to assist. A tow line was connected to the yacht and the weight was taken while the anchor was manually recovered. Once the anchor was safely stowed the yacht was towed into Falmouth Inner Harbour where it was placed on a mooring, the lifeboat being released to return to its station at 5.32pm.

The all-weather lifeboat was back alongside its pontoon berth by 5.40pm where it was refuelled and made ready for service by 5.50pm.

The yacht had been anchored in Falmouth Bay for several days. She was having mechanical issues and with the increasing South westerly winds there were concerns in respect to the crew's wellbeing. A rigid inflatable from Falmouth Boat Construction had also proceeded to assist the yacht and their help in the service was very much appreciated.

Small open boat broken down

Date: Saturday 8 August 2020

Location: 2 miles East of Manacle Buoy

Weather: N 4-5, moderate sea and slight swell, excellent visibility, clear sky

Inshore lifeboat: (launched 3.18pm/ returned 5.27pm) – Nick Head, (Helmsman), Tamsin Mulcahy, Cadan Harris

Details:

At 3.12pm Falmouth Coastguard requested that Falmouth inshore lifeboat be launched to assist a 4.8m open boat with one person on board which had suffered engine failure while fishing in Falmouth Bay.

The inshore lifeboat launched from her slipway at 3.18pm and headed out into Falmouth Bay to try and locate the casualty. The boat was finally located 2 miles East of Manacle Buoy with the inshore lifeboat arriving on scene at 3.48pm. A tow was established and the boat was towed back to Falmouth and up the Penryn River where it was safely moored alongside the Falmouth Marina at 5.14pm, the inshore lifeboat then being released to return to its station.

The inshore lifeboat was recovered onto its slipway at 5.27pm where the lifeboat was refuelled and made ready for service by 6.00pm.

The casualty had been fishing in Falmouth Bay when the engine had failed, so used a mobile phone to call the coastguard for assistance. Being unsure of its position made the task of locating the boat more difficult but it was eventually located over five miles south of Falmouth having been unable to anchor due to the depth of water and drifting further out to sea in the strong offshore winds.

Sailing boat aground on rocks

Date: Sunday 9 August 2020

Location: Pendennis Point

Weather: N 4-5, slight sea and no swell, fair visibility, overcast

Inshore lifeboat: (launched 7.17pm/ returned 8.18pm) – Nick Head, (Helmsman), Tom Bird, Tamara Brookes, Andy Edwards

Details:

At 7.09pm Falmouth Coastguard requested that Falmouth inshore lifeboat be launched to assist a 6m sailing boat with three people on board which had reported being aground on Pendennis Point. The Falmouth Coastguard Cliff Rescue Team was also tasked to assist.

The inshore lifeboat launched from her slipway at 7.17pm arriving on scene at 7.22pm. The sailing boat had been towed clear of the rocks by another yacht and was between Pendennis Point and Black Rock. Having checked that the sailing boat was not taking any water, it was taken in tow by the inshore lifeboat at 7.30pm. The boat was taken back into the inner harbour and then up the Penryn River to its mooring off Trevissome. Once the boat was safely on its mooring at 8.04pm, the inshore lifeboat was released to return to its station, while the casualties rowed ashore to be met by the coastguard team.

The inshore lifeboat was recovered onto its slipway at 8.18pm where the lifeboat was refuelled and made ready for service by 8.45pm.

The sailing boat was motoring using its small outboard when on rounding Pendennis Point they found that it could not make any headway against the strong northerly wind and ended up on the rocks. They did not consider using their sails and suitable safety advice was given to the casualties by both the lifeboat crew and the coastguard team.

Medical Evacuation from yacht

Date: Tuesday 11 August 2020

Location: Percuil River

Weather: N 3, slight sea and no swell, poor visibility, overcast with mist

Inshore lifeboat: (launched 5.42am/ returned 7.50am) – Jamie Wakefield, (Helmsman), Tamara Brookes, Chris Simpson, Joe Amps

Details:

At 5.46am Falmouth Coastguard requested that Falmouth inshore lifeboat be launched following a call from the Ambulance service requesting the medical evacuation of the sole occupant of a yacht which was at anchor in the Percuil River.

The inshore lifeboat launched from her slipway at 5.42am and headed up the Percuil River arriving on scene at 6.11am. Two lifeboat crew trained in casualty care went on board the yacht to assess the casualty's condition before the casualty was transferred to the inshore lifeboat. It was agreed with Ambulance control that the casualty would be taken back to Falmouth where they would arrange for an ambulance to meet them, so at 5.30am the inshore lifeboat started heading back down the river.

The inshore lifeboat arrived back at its station at 6.54am and moored alongside the all-weather lifeboat to allow the casualty to be transferred ashore and be taken to the boathouse to await the arrival of the Ambulance. The Ambulance arrived at 7.13am and the casualty was handed over to the paramedics.

The inshore lifeboat was recovered onto its slipway at 7.50am where the lifeboat was refuelled and made ready for service by 8.24am.

Motorboat broken down

Date: Tuesday 11 August 2020

Location: Off Mylor, Carrick Roads

Weather: N 3, slight sea and no swell, poor visibility, overcast with mist

Inshore lifeboat: (launched 9.30pm/ returned 10.33pm) – Jamie Wakefield, (Helmsman), Tom Bird, Neil Capper, Tamsin Mulcahy

Details:

At 9.22pm Falmouth Coastguard requested that Falmouth inshore lifeboat be launched following a 999 call from a 6.5m motorboat with two people on board which had broken down off Mylor in the Carrick Roads.

The inshore lifeboat launched from her slipway at 9.30pm and headed up the Carrick Roads locating the vessel between Mylor and the Carrick Buoy at 9.39pm. Once the boats anchor had been recovered at 9.43pm it was taken in tow back to Loe Beach where it was safely placed on its trailer by 10.14pm, the inshore lifeboat then being released to return to its station.

The inshore lifeboat was recovered onto its slipway at 10.33pm where the lifeboat was refuelled and made ready for service by 11.02pm.

The motorboat had suffered engine failure while out fishing in the Carrick Roads. The occupants had deployed the boats anchor and contacted the Coastguard for assistance using their mobile phone. Suitable safety advice was given to the casualties as they did not have any lifejackets or a VHF radio.

Rigid inflatable broken down

Date: Thursday 13 August 2020

Location: Falmouth Bay

Weather: WSW 2, slight sea and no swell, fair visibility, partly cloudy

Inshore lifeboat: (launched 5.58pm/ returned 6.48pm) – Jamie Wakefield, (Helmsman), Luke Wills, Tom Bird, Chris Simpson

Details:

At 5.50pm Falmouth Coastguard requested that Falmouth inshore lifeboat be launched following a 999 call from a 6.5m rigid inflatable with two people on board which had broken down in Falmouth Bay. The boat was unsure of its position but believed they were about 3 miles offshore.

The inshore lifeboat launched from her slipway at 5.58pm and headed towards St Anthony Lighthouse before searching seawards into Falmouth Bay. At 6.12pm the rigid inflatable was located 1.5 miles South of St Anthony Lighthouse. A towline was passed to the boat and at 6.16pm and the tow back into Falmouth commenced. Once in Falmouth Inner Harbour the boat was taken to the Grove Place Dinghy Park where it was safely moored up at 6.45pm so the inshore lifeboat was then released to return to its station.

The inshore lifeboat was recovered onto its slipway at 6.48pm where the lifeboat was refuelled and made ready for service by 6.59pm.

The rigid inflatable had suffered engine failure due to a fuel problem. They had initially tried to get another vessel to come out to assist them, but this was delayed, so with an uncertain location and reducing visibility it was decided to task the inshore lifeboat to assist them. The boat did not have a GPS, charts or a working VHF radio all of which would have helped locate the vessel so suitable safety advice was given, but its occupants did have lifejackets and had tried to resolve the situation themselves.

Search for missing swimmer

Date: Saturday 15 August 2020

Location: off Pennance Point, Falmouth Bay

Weather: SW 3, slight sea and no swell, excellent visibility, sunny periods

All-weather lifeboat: (launched 11.02am / returned 11.07am) – Luke Wills (2nd Coxswain), Dave Nicoll, Tom Bird, Adam West, Will Allen, Andy Edwards, Chris Simpson, Jemima Henstridge-Blows

Inshore lifeboat: launched 10.43am / returned 11.24am) – Jamie Wakefield (Helmsman), Joe Amps, Jack Williams, James Murray

Details:

At 10.33am Falmouth Coastguard requested that Falmouth inshore lifeboat be launched following a report of a missing swimmer between Swanpool and Pennance Point. The Falmouth Coastguard Cliff Rescue Team was also tasked to assist.

The inshore lifeboat launched from her slipway at 10.43am and having rounded Pendennis Point headed across the bay arriving on scene at Swanpool at 10.48am and commenced a search from Swanpool Beach to Pennance Point. At 10.54am the first informant was located on a kayak who advised that the swimmer was last seen heading out to sea so the inshore lifeboat immediately requested the launch of the all-weather lifeboat to assist in the search. The inshore lifeboat continued its search to Pennance Point and then started an expanding square search off Pennance Point.

The all-weather lifeboat was paged at 10.55am and left its pontoon berth at 11.02am. At 11.04am it was confirmed that the missing swimmer had been located by the Coastguard Team at Maenporth Beach and was safe and well so both lifeboats were stood down to return to their station.

The all-weather lifeboat was back alongside its pontoon berth at 11.07am and was made ready for service by 11.15am. The inshore lifeboat left the scene at 11.09am and was back on its slipway by 11.24am where it was refuelled and made ready for service by 11.55am.

The swimmer who was wearing a brightly coloured swim cap and towing a swim buoy was being monitored from the shore by one of their family. When they were having difficulty spotting the swimmer the family member hired a Kayak to follow them but then lost sight of the swimmer off Pennance Point so wisely contacted the emergency services using a mobile phone to raise the alarm. The inshore lifeboat had been assisted in its initial search by a rigid inflatable from the Elemental Watersports Centre and also one of the lifeboat crew who was afloat in their own boat.

Report of an unmanned open boat

Date: Saturday 15 August 2020

Location: off Greeb Point, Gerrans Bay

Weather: SW 3, slight sea and no swell, good visibility, sunny periods

All-weather lifeboat: (launched 3.36pm / returned 4.07pm) – Luke Wills (2nd Coxswain), Dave Nicoll, Adam West, Will Allen, Andy Edwards, Joe Amps, Jack Williams

Inshore lifeboat: (launched 3.32pm / returned 4.08pm) – Jamie Wakefield (Helmsman), Tom Bird, Neil Capper, James Murray

Details:

At 3.24pm Falmouth Coastguard requested that Falmouth inshore lifeboat be launched following a report from the Portscatho NCI Lookout of an unmanned open boat adrift off Greeb Point in Gerrans Bay. The decision was made at 3.30pm to also launch the all-weather lifeboat to provide additional resource should a search be required.

The inshore lifeboat launched from her slipway at 3.32pm followed by the all-weather lifeboat which left its pontoon berth at 3.36pm. The inshore lifeboat rounded Zone Point and headed up to the east arriving on scene at 3.44pm. The open boat was located at anchor and by 3.50pm the inshore lifeboat crew had located the owner on the shoreline who confirmed all was well so both lifeboat were stood down to return to their station.

The all-weather lifeboat was back alongside its pontoon berth at 4.07pm and was made ready for service by 4.20pm. The inshore lifeboat was back on its slipway by 4.08pm where it was refuelled and made ready for service by 4.33pm.

The Portscatho NCI Lookout had become concerned for the wellbeing of the occupants of a 5m open boat with its engine down and fishing rods rigged which appeared to be drifting with no one on board off Greeb Point. The occupants had safely anchored their boat and gone ashore on kayaks so while this turned out to be a false alarm the situation could have been totally different.

Yacht with rudder failure

Date: Wednesday 19 August 2020

Location: Falmouth Bay

Weather: SSW 5-6, moderate sea and swell, poor visibility, overcast with rain

All-weather lifeboat: (launched 10.01am / returned 11.43am) – Jonathon Blakeston (Coxswain), Luke Wills, Carl Beardmore, Dave Nicoll, Tom Bird, Jamie Wakefield, Will Allen

Details:

At 9.53am Falmouth Coastguard requested that Falmouth all-weather lifeboat be launched to assist a 9m yacht with one person on board which had suffered rudder failure in Falmouth Bay.

The all-weather lifeboat left its pontoon berth at 10.01am and having rounded Pendennis Point set course for the casualty which was 1 mile South West of Pendennis Point arriving on scene at 10.13am. Having assessed the situation a drogue was passed to the casualty to help stabilise it during the tow. At 10.27am the tow line had been connected and the tow into Falmouth could begin. Once in the shelter of the Docks the yacht was brought alongside the lifeboat before being taken up the Penryn River to Coastlines Wharf where the yacht was transferred to the workboat from Falmouth Premier Marina, the all-weather lifeboat being released at 11.31am to return to its station.

The all-weather lifeboat was back alongside its pontoon berth by 11.43am where it was refuelled and made ready for service by 12.00pm.

The yacht was on passage around the United Kingdom and having been moored in the Helford was heading for Falmouth to shelter ahead of the forecast gales when the rudder had failed.

Reports of multiple people caught in rip current

Date: Thursday 20 August 2020

Location: Maenporth Beach

Weather: S 6-7, rough sea and moderate swell, fair visibility, partly cloudy

Inshore lifeboat: (launched 3.24pm/ returned 4.53pm) – Jamie Wakefield, (Helmsman), Tom Bird, Jack Williams, James Murray

Details:

At 3.19pm Falmouth Coastguard requested that Falmouth inshore lifeboat be launched following reports of multiple people being caught in the rip current at Maenporth Beach. Falmouth and Porthoustock Coastguard Rescue Teams were also tasked to assist.

The inshore lifeboat launched from her slipway at 3.24pm and having rounded Pendennis Point made best speed in the rough conditions arriving on scene off Maenporth Beach at 3.35pm. It was soon established that all those involved had been assisted ashore by a member of the public on a kayak. Once the coastguard team were happy with the situation at 3.45pm the inshore lifeboat was released to return to its station.

As the inshore lifeboat was passing the docks at 3.54pm and heading back into the harbour there were reports of more people in difficulty at Maenporth so the inshore lifeboat was requested to return to the scene arriving back at 4.01pm. The additional casualties had been assisted ashore by a Coastguard Rescue swimmer, but the inshore lifeboat was requested to remain on scene while the Coastguard ensured no one else got into difficulty. At 4.40pm the inshore lifeboat was again released to return to its station.

The inshore lifeboat was recovered onto its slipway at 4.53pm where the lifeboat was refuelled and made ready for service by 5.14pm.

In total eleven people were assisted, eight by an off duty member of the North Dartmoor Search and Rescue Team on his kayak and a further three were brought ashore by the Coastguard Rescue swimmer. Maenporth Beach does not have a lifeguard service and has significant rip currents when there are strong south or easterly conditions.

Paddle Boarders reported in difficulties

Date: Friday 21 August 2020

Location: Chapel Point, St Austell Bay

Weather: SW 5-7, moderate sea and swell, good visibility, partly cloudy

All-weather lifeboat: (launched 6.40pm / returned 7.42pm) – Jonathon Blakeston (Coxswain), Luke Wills, Dave Nicoll, Tom Bird, Andy Jenkin, Jamie Wakefield, Jack Williams

Details:

At 6.30pm Falmouth Coastguard requested that Falmouth all-weather lifeboat be launched to support the Fowey lifeboat which was responding to a report of two paddle boarders in difficulties off Chapel Point in St Austell Bay. The Mevagissey Coastguard Rescue Team was also tasked to assist.

The all-weather lifeboat left its pontoon berth at 6.40pm and having rounded Zone Point headed up to the east. At 7.09pm when the lifeboat was 3 miles South West of Dodman Point it was confirmed that the two paddle boarders had made it ashore safely, so the lifeboat was released to return to its station.

The all-weather lifeboat was back alongside its pontoon berth by 7.42pm where it was refuelled and made ready for service by 8.00pm.

Inflatable Kayak being blown out to sea

Date: Sunday 23 August 2020

Location: Gyllyngvase Beach

Weather: WSW 4, slight sea and swell, excellent visibility, partly cloudy

Inshore lifeboat: (launched 12.48pm/ returned 1.05pm) – Jamie Connolly (Helmsman), Chris Simpson, Cadan Harris, Jemima Henstridge-Blows

Details:

At 12.34pm Falmouth Coastguard requested that Falmouth inshore lifeboat be launched following reports from the Gyllyngvase Lifeguards and members of the public of an inflatable kayak with two persons on board being blown towards St Anthony Lighthouse.

The inshore lifeboat launched from her slipway at 12.48pm arrived on scene locating the kayak off Pendennis Point at 12.55pm. The two occupants were rowing towards Pendennis Point and once they were approached Crab Quay at 1.00pm the inshore lifeboat was released to return to its station.

The inshore lifeboat was recovered onto its slipway at 1.05pm where the lifeboat was refuelled and made ready for service by 1.15pm.

The two occupants of the inflatable kayak had launched the craft from Gyllyngvase Beach with the intention of rowing to Swanpool Beach. When they ended up being blown towards Pendennis

Point, they had managed to paddle into Crab Quay where they landed safely, presumably deciding to walk the 1.5 miles back to Gyllyngvase Beach.

Medical evacuation of person with ankle injury

Date: Wednesday 26 August 2020

Location: Pennance Point

Weather: W 4, slight sea and swell, excellent visibility, partly cloudy

Inshore lifeboat: (launched 1.05pm/ returned 2.45pm) – Jamie Wakefield (Helmsman), Tamara Brookes, Jack Williams, Lloyd Barron

Details:

At 12.59pm Falmouth Coastguard requested that Falmouth inshore lifeboat be launched to assist in the medical evacuation of a casualty with an ankle injury at Pennance Point. The Falmouth and Porthoustock Coastguard Rescue Teams and paramedics were also tasked to assist.

The inshore lifeboat launched from her slipway at 1.05pm and having rounded Pendennis Point headed across the bay arriving on scene at Pennance Point at 1.14pm. Two casualty care trained crew members were placed on the rocks to assess and treat the casualty. As access to the scene was difficult it was decided to evacuate the casualty using the inshore lifeboat. The inshore lifeboat proceeded to Swanpool Beach where it collected a paramedic and stretcher and transferred them to the scene. Once the casualty had been given pain relief they were placed in a stretcher and transferred to the inshore lifeboat, accompanied by the paramedic and three members of the Coastguard Team. They were taken to Swanpool Beach where the casualty and emergency personnel were landed safely, the inshore lifeboat being released to return to its station at 2.31pm.

The inshore lifeboat was recovered onto its slipway at 2.45pm where the lifeboat was refuelled and made ready for service by 3.22pm.

The casualty had been exploring the gulleys and rocks around the headland when they had fallen from a cliff and injured their ankle. Access to the location was very restricted so it was decided the best option was to evacuate the casualty using the inshore lifeboat. This was an excellent example of the various emergency services working together to achieve a successful outcome.

Report of dinghy adrift

Date: Thursday 27 August 2020

Location: upstream from Greenbank Quay, Penryn River

Weather: SW 5, slight sea and no swell, poor visibility, overcast with heavy rain

Inshore lifeboat: (launched 11.22am/ returned 11.54am) – Jamie Wakefield (Helmsman), Tamara Brookes, Tamsin Mulcahy, Lloyd Barron

Details:

At 11.08am Falmouth Coastguard requested that Falmouth inshore lifeboat be launched following reports of a dinghy adrift with its engine down in the Penryn River above Greenbank Quay. The Falmouth Coastguard Rescue Team was also tasked to assist.

The inshore lifeboat launched from her slipway at 11.22am and headed up the Inner Harbour arriving off the Greenbank Quay at the start of the Penryn River and at 11.33am the dinghy was located on the shoreline close to Coastlines Wharf. A lifeboat crewmember was placed ashore to check for any recent occupancy. With the engine being cold and no signs of anyone having been on board, the dinghy was recovered from the shoreline at 11.42am and towed back to the lifeboat station where it was safely secured to the station's pontoon at 11.50am.

The inshore lifeboat was recovered onto its slipway at 11.54am where the lifeboat was refuelled and made ready for service by 12.32pm.

Later in the afternoon the owner visited the station and was reunited with their 3m rigid inflatable dinghy, thanking the crew for its safe recovery.

Cabin boat adrift

Date: Sunday 30 August 2020

Location: off Kiln Quay, Falmouth Inner Harbour

Weather: NE 2, slight sea and no swell, excellent visibility, partly cloudy

Inshore lifeboat: (launched 11.15am / returned 11.56am) – Jamie Connolly (Helmsman), Andy Edwards, Chris Simpson, Jemima Henstridge-Blows

Details:

At 11.11am Falmouth Coastguard requested that Falmouth inshore lifeboat be launched following a report of a cabin boat adrift off Kiln Quay in Falmouth Inner Harbour. The first informant reported that the outboard engine was down, and the kill cord was in place raising concerns that someone might be in the water. The Falmouth Coastguard Rescue Team was also tasked to assist.

The inshore lifeboat launched from her slipway at 11.15am and headed across the Inner Harbour arriving on scene in the moorings between Kiln Quay and Flushing at 11.18am. Having carried out an initial check of the cabin boat, the inshore lifeboat was then tasked to commence a search from Falmouth Inner Harbour to Penarrow Point in the Carrick Roads. The Coastguard Team went to Trefusis Point to carry out a visual check from the high ground.

The Coastguard meanwhile were carrying out checks to try and identify the owner and their safety. At 11.36am they were able to confirm the mooring that was associated with the boat's owner, so the inshore lifeboat was diverted to check this mooring which was found to be empty. At 11.39am it was confirmed that the owner was safe and well, so the inshore lifeboat placed the cabin boat back on its mooring before at 11.50am being released to return to its station.

The inshore lifeboat was recovered onto its slipway at 11.56am where the lifeboat was refuelled and made ready for service by 12.15pm.

The owner of the cabin boat was afloat racing on his other boat and confirmed that the boat had been used earlier and had been left on the mooring. The first informant had done the right thing in notifying the Coastguard as things could have been totally different with potentially a person in the water.

Emergency Position Indicating Radio Beacon alert

Date: Friday 11 September 2020

Location: Falmouth Harbour

Weather: S 2-3, calm sea and no swell, good visibility, overcast

Inshore lifeboat: (launched 8.32am / returned 9.03am) – Jamie Connolly (Helmsman), Tom Bird, Tamara Brookes, Lloyd Barron

Details:

At 8.21am Falmouth Coastguard requested that Falmouth inshore lifeboat be launched following a distress alert from an EPIRB (Emergency Position Indicating Radio Beacon) with reported positions in both Falmouth Harbour and the Penryn River. The Falmouth Coastguard Rescue Team was also tasked to assist.

The inshore lifeboat launched from her slipway at 8.32am and was requested to proceed to a mooring in the Inner Harbour where it was believed the yacht which the beacon was registered to had been moored. At 8.40am the yacht was located on the mooring and it was confirmed that it was unoccupied and secure. Falmouth Coastguard managed to contact the yachts owner who gave permission for the lifeboat crew to enter the yacht where they located the beacon. The yacht was secured and at 8.58am the inshore lifeboat headed back to its station with the beacon which was handed into the care of the Coastguard Rescue Team.

The inshore lifeboat was recovered onto its slipway at 9.03am where the lifeboat was refuelled and made ready for service by 9.20am.

The beacon had self-activated and had been picked up by the satellite system with the alert being passed to Falmouth Coastguard. It had proved impossible to deactivate the beacon, so it was handed over to the Coastguard to sort.

Yacht dragging its anchor

Date: Thursday 17 September 2020

Location: St Mawes Harbour

Weather: E 6-7, moderate sea and slight swell, good visibility, partly cloudy

Inshore lifeboat: (launched 5.15am / returned 5.54am) – Jamie Wakefield (Helmsman), Tom Bird, Tamara Brookes, Jack Williams

Details:

At 5.06am Falmouth Coastguard requested that Falmouth inshore lifeboat be launched following a Mayday call from an 11m yacht with two persons on board which was dragging its anchor near St Mawes Harbour.

The inshore lifeboat launched from her slipway at 5.15am and located the yacht off Tavern Beach at 5.23am. Once on scene it was found that the skipper of the yacht had managed to pick up the pennant of a nearby mooring which had helped to stop it from dragging ashore. The inshore lifeboat helped to take the weight of the yacht while its crew secured it to the adjacent mooring. Once the skipper confirmed he was happy with the situation the inshore lifeboat was at 5.42am released to return to its station.

The inshore lifeboat was recovered onto its slipway at 5.54am where the lifeboat was refuelled and made ready for service by 6.06am.

The skipper of the yacht had issued a Mayday when it had started dragging its anchor towards the shore and its engine had failed to start. Luckily the yachts crew had managed to pick up the pennant from a nearby mooring but with the yachts anchor still deployed it was decided that the best option was for the inshore lifeboat to assist in securing the yacht to the mooring so that the skipper could arrange for recovery in daylight. The occupants of the yacht were very grateful for the assistance given and had been clearly shaken by the situation.

Medical evacuation from a yacht

Date: Friday 18 September 2020

Location: Helford River

Weather: E 6, moderate sea and heavy swell, good visibility, partly cloudy

All-weather lifeboat: (launched 4.02pm / returned 5.22pm) – Jonathon Blakeston (Coxswain), Luke Wills, Dave Nicoll, Tom Bird, Tamara Brookes, James Murray, Lloyd Barron

Details:

At 3.50pm Falmouth Coastguard requested that Falmouth all-weather lifeboat be launched to evacuate a sailor on an 11m yacht who had broken his forearm. The yacht had been off the entrance to the Helford River and was heading up the river to find some shelter from heavy easterly swell.

The all-weather lifeboat left its pontoon berth at 4.02pm and having rounded Pendennis Point headed across Falmouth Bay, locating the yacht on a visitors mooring off Helford Passage at 4.27pm. Two lifeboat crew members trained in casualty care were transferred to the yacht to assess the casualty. They placed the casualty's arm in a temporary splint before transferring him to the all-weather lifeboat which departed the Helford at 4.46pm and headed back to Falmouth where an ambulance had been requested to meet the lifeboat on its arrival at its station.

The all-weather lifeboat was back alongside its pontoon berth by 5.22pm where the casualty was transferred ashore. The ambulance had been diverted to another incident, so arrangements were made for the casualty to be transferred by road to the A&E Department at Treliske Hospital. The lifeboat was refuelled and made ready for service by 5.50pm.

Swimmer in difficulties

Date: Friday 18 September 2020

Location: Gyllyngvase Beach

Weather: E 5, moderate sea and heavy swell, good visibility, partly cloudy

Inshore lifeboat: (launched 7.09pm / returned 7.38pm) – Jamie Wakefield (Helmsman), Tom Bird, Tamara Brookes, James Murray

Details:

At 7.03pm Falmouth Coastguard requested that Falmouth inshore lifeboat be launched following reports of a person in difficulty in the water at Gyllyngvase Beach. The Falmouth Coastguard Rescue Team and the Coastguard helicopter were also tasked to assist. Further reports were received reporting three other people in difficulty so the all-weather lifeboat was also requested to launch at 7.08pm, fortunately by 7.11pm these casualties were confirmed as safe ashore.

The inshore lifeboat launched from her slipway at 7.09pm and having rounded Pendennis Point headed along the seafront towards Gyllyngvase Beach arriving on scene at 7.14pm. The inshore lifeboat commenced an initial search of the area but the Coastguard Team were soon able to confirm that the initial casualty had been assisted ashore so at 7.19pm the inshore lifeboat was released to return to its station.

The inshore lifeboat arrived back at its station at 7.32pm but was asked to remain afloat while the Coastguard Team investigate a report of a missing surfer, but by 7.38pm the surfer had been located ashore so the inshore lifeboat was recovered onto its slipway where it was refuelled and made ready for service by 7.54pm.

The strong easterly winds had created a heavy swell at both Gyllyngvase and Swanpool Beaches which had led to several swimmers and surfers getting into difficulties.

Yacht with engine failure and becalmed

Date: Saturday 20 September 2020

Location: 1.5 miles East of Gull Rock

Weather: N 2, slight sea and swell, good visibility, clear sky

All-weather lifeboat: (launched 7.28pm / returned 10.25pm) – Jonathon Blakeston (Coxswain), Dave Nicoll, Tom Bird, Jonathan Hackwell, Sandy Procter, Neil Capper, Nick Head, Andy Edwards

Details:

At 7.14pm Falmouth Coastguard requested that Falmouth all-weather lifeboat be launched to assist a 11m yacht with two persons on board which had suffered engine failure and was becalmed in Veryan Bay, 1.5 miles East of Gull Rock.

The all-weather lifeboat left its pontoon berth at 7.28pm and having rounded Zone Point headed east arriving on scene at 8.03pm. A line was passed to the yacht and by 8.10pm the tow back to Falmouth had commenced. The yacht was towed to the Carrick Roads where the tow was shortened and the yacht was then taken alongside the lifeboat. It was then placed alongside another yacht which was anchored close the St Just Pool so that repairs to the engine could take place. At 10.05pm the lifeboat was released to return to its station.

The all-weather lifeboat was back alongside its pontoon berth by 10.25pm where it was refuelled and made ready for service by 10.45pm.

The yacht had put out a Pan Pan broadcast but when no other vessels were able to assist, and the light was fading it was decided to launch the lifeboat. The crew of the yacht were very grateful for the assistance they were given.

Sailing boat reported adrift

Date: Thursday 24 September 2020

Location: Falmouth Bay

Weather: NW 5, Moderate Sea and Slight Swell, Poor Visibility, Overcast with Rain and Hail

Inshore lifeboat: (Launched 18:05 / Returned 19:14) – Jamie Connolly (Helmsman), Adam West, Tamsin Mulcahy, Jack Williams

Details:

At 17:39 Falmouth Coastguard requested that Falmouth Inshore Lifeboat be launched following a report from the Bunker Barge Lizrix of a small sailing boat adrift in Falmouth Bay.

The Inshore Lifeboat launched from her slipway at 17:52 and having rounded Pendennis Point headed towards the bunker barge arriving on scene at 18:05. With visibility being poor due to the heavy rain and hail the crew of the barge helped direct the Inshore Lifeboat and at 18:09 the small sailing boat had been located 1 mile North East of Nare Point. A lifeboat crew member was placed on board the boat and was able to confirm that there was no sign of recent occupancy so a line was connected to the boat and it was towed into the Helford where it was placed on a visitors mooring at 19:01, the Inshore Lifeboat being released to return to its station.

The Inshore Lifeboat was recovered onto its slipway at 19:14 where it was refuelled and made ready for service by 19:34.

It appears that the sailing boat had parted its mooring and drifted out of the Helford on the ebb tide and north westerly wind. Shortly after the inshore lifeboat had returned to its station the coastguard advised that they had managed to contact the owner who was grateful for the recovery of the boat.

Yacht in difficulties due to fouled spinnaker

Date: Saturday 10 October 2020

Location: 3 miles South of Pendennis Point

Weather: NNW 5-6, moderate sea and slight swell, good visibility, partly cloudy

All-weather lifeboat: (launched 2.29pm / returned 4.14pm) – Dave Nicoll (Deputy 2nd Coxswain), Carl Beardmore, Jonathan Hackwell, Nick Head, Tamsin Mulcahy, Chris Simpson, Cadan Harris

Inshore lifeboat: (launched 2.17pm / returned 4.20pm) – Elliot Holman (Helmsman), Tom Bird, Neil Capper, Adam West

Details:

At 2.07pm Falmouth Coastguard requested that Falmouth inshore lifeboat be launched to assist a 10m racing yacht which had broached, and its spinnaker had become fouled off Pendennis Point. Other yachts in the area were trying to assist.

The inshore lifeboat launched from her slipway at 2.17pm and having rounded Pendennis Point headed south to locate the yacht. At 2.23pm the decision was taken to also launch the all-weather lifeboat in case additional assistance was required, the all-weather lifeboat leaving its pontoon berth at 2.29pm. The inshore lifeboat located the yacht 3 miles South of Pendennis Point at 2.29pm having been blown downwind with its the spinnaker still wrapped around its forestay. The inshore lifeboat placed a crew member on board the yacht and a tow line was connected by 2.39pm, the all-weather lifeboat arriving on scene at 2.42pm.

The inshore lifeboat towed the yacht back towards Falmouth escorted by the all-weather lifeboat. Once back in the harbour under the lee of Trefusis Point one of the yacht's crew was able to go up the mast to cut the spinnaker free before the inshore lifeboat towed the yacht to its mooring in the Inner Harbour. The all-weather lifeboat was released to return to its station at 4.06pm. Once the yacht was safely on its mooring at 4.12pm the inshore lifeboat was also released to return to its station.

The all-weather lifeboat was back alongside its pontoon berth at 4.14pm and the inshore lifeboat was back on its slipway by 4.20pm. Both lifeboats were refuelled and made ready for service by 4.35pm.

The yacht had been taking part in a race off Pendennis Point when it had broached, the spinnaker had got wrapped around the forestay and one of its crew had fallen overboard. A nearby yacht recovered the man overboard while the three remaining crew tried to drop the spinnaker, but this proved impossible as it was caught high up in the rigging and in the strong winds the yacht was blown out into the bay.

Search for missing snorkeler

Date: Saturday 10 October 2020

Location: area of Gyllyngvase and Swanpool Beaches

Weather: WNW 4, slight sea and swell, good visibility, partly cloudy with showers

All-weather lifeboat: lLaunched 5.30pm / returned 6.10pm) – Dave Nicoll (Deputy 2nd Coxswain), Carl Beardmore, Jonathan Hackwell, Will Allen, Tamsin Mulcahy, Chris Simpson, Cadan Harris

Inshore lifeboat: (launched 5.30pm / Returned 6.12pm) – Elliot Holman (Helmsman), Tom Bird, Neil Capper, Adam West

Details:

At 5.24pm Falmouth Coastguard requested that Falmouth all-weather and inshore lifeboats be launched to assist in a search for a snorkeler who had been reported overdue in the area of Gyllyngvase and Swanpool Beaches. The Falmouth Coastguard Rescue Team and Police were also tasked to assist in the search.

The inshore lifeboat launched from her slipway and the all-weather lfeboat which left its pontoon berth, both at 5.30pm. The inshore lifeboat commenced its search from Pendennis Point at 5.38pm, with the all-weather lifeboat heading across the bay to Pennance Point where it started its search towards Swanpool Beach at 5.40pm. Both lifeboats were off Gyllyngvase Beach at 5.50pm having completed their initial searches and were about to extend their searches when at 5.55pm the Coastguard advised that further information had been received and all search and rescue units were released to return to their stations.

The all-weather lifeboat was back alongside its pontoon berth at 6.10pm and the inshore lifeboat was back on its slipway by 6.12pm. Both lifeboats were refuelled and made ready for service by 6.30pm.

Medical evacuation from yacht

Date: Sunday 11 October 2020

Location: off Grebe Beach, near Durgan in Helford Estuary

Weather: NW 4, slight sea and swell, excellent visibility, partly cloudy

All-weather lifeboat: (launched 2.55pm / returned 6.00pm) – Dave Nicoll (Deputy 2nd Coxswain), Tom Bird, Andy Jenkin, Jonathan Hackwell, Sandy Procter, Neil Capper, Will Allen, Andy Edwards

Details:

At 2.43pm Falmouth Coastguard requested that Falmouth all-weather lifeboat be launched to assist with the medical evacuation of a person from a 11m yacht with two persons on board which was anchored off Grebe Beach, near Durgan in the Helford Estuary. Falmouth Coastguard Rescue Team, the Coastguard helicopter and an ambulance were also tasked to assist.

The all-weather lifeboat left its pontoon berth at 2.55pm and having rounded Pendennis Point headed across the bay towards the Helford Estuary arriving off Dugan at 3.17pm. Another yacht had moored up alongside the casualty vessel to provide assistance. The Coastguard helicopter which had been airborne on a routine exercise was already on scene and lowered its winchman paramedic to the yacht. The helicopter requested that the lifeboat go alongside the yacht to provide additional assistance as required, the helicopter relocating to a landing spot in a nearby field which had been prepared by the Coastguard Team.

Once the casualty had been assessed it was decided that the best option would be to transfer them ashore to the ambulance using the lifeboats inflatable Y-boat. The casualty and winchman paramedic were taken ashore to Durgan where they were safely landed at 3.50pm, the casualty being taken to hospital by ambulance while the winchman was taken by the coastguard team to the waiting helicopter.

Having returned to the lifeboat, two crew members were put on board the yacht to help take it back to its mooring at Mylor, so that its remaining occupant could join the casualty at hospital. The yachts anchor was recovered by 4.00pm and it headed out of the Helford and across Falmouth Bay escorted by the all-weather lifeboat. The yacht was secured safely on its mooring and the occupant assisted ashore by 5.45pm, the all-weather lifeboat then being released to return to its station.

The all-weather lifeboat was back alongside its pontoon berth by 6.00pm where it was refuelled and made ready for service by 6.30pm.

The casualty had become unwell on board their yacht while it was anchored in the Helford. Due to the casualties condition the winchman paramedic decided that the patient would be best to be transferred to hospital by ambulance allowing them to be monitored on route.

Capsized kayaks with people in the water

Date: Thursday 22 October 2020

Location: Bass Point, The Lizard

Weather: N 3, slight sea and swell, excellent visibility, partly cloudy

All-weather lifeboat: (launched 1.14pm / returned 3.28pm) – Jonathon Blakeston (Coxswain), Carl Beardmore, Dave Nicoll, Tom Bird, Will Allen, Lloyd Barron

Details:

At 1.06pm Falmouth Coastguard requested that Falmouth all-weather lifeboat be launched to assist two people and a dog who were reported to be in the water after the capsize of their kayaks off Bass Point near The Lizard. The casualties had been spotted by the Bass Point NCI Lookout and Mullion Coastguard Rescue Team was also tasked to assist. The Falmouth lifeboat was tasked as The Lizard lifeboat was already out on a search for an overturned boat 25 miles SSW of Lizard Point.

The all-weather lifeboat left its pontoon berth at 1.14pm and having rounded Pendennis Point headed south past the Manacles and Blackhead arriving on scene at 1.52pm. It was confirmed that one of the casualties and the dog had managed to reach the rocks below Bass Point and was able to climb to safety. The other casualty was recovered from the water by a local fishing boat and was taken to The Lizard lifeboat slipway where they were landed. Once it was confirmed that all casualties were safe the Falmouth Lifeboat carried out a search for the remaining Kayak which was recovered south of Bass Point, with the paddle being found close under the headland. These items were taken back to the lifeboat slipway and landed ashore before the lifeboat was released at 2.35pm to return to its station.

The all-weather lifeboat was back alongside its pontoon berth by 3.28pm where it was refuelled and made ready for service by 3.45pm.

The casualties were not injured and were grateful for the assistance they were given. Without the help of the NCI Lookout, the prompt response of the local fishing boat and the support provided by Mullion Coastguard Team the outcome of this incident could have been very different.

Windsurfer in difficulties

Date: Saturday 24 October 2020

Location: between Castle and Gyllyngvase Beaches

Weather: SW 3-4, moderate sea and swell, fair visibility, overcast with rain

Inshore lifeboat: (launched 10.07am / returned 10.38am) – Nick Head (Helmsman), Tom Bird, Jack Williams, Cadan Harris

Details:

At 9.58am Falmouth Coastguard requested that Falmouth inshore lifeboat be launched following several reports of a windsurfer in difficulties between Castle and Gyllyngvase Beaches.

The inshore lifeboat launched from her slipway at 10.07am and having rounded Pendennis Point headed towards the last reported position of the windsurfer. Arriving on scene at 10.14am the windsurfer was located and was taken on board the inshore lifeboat along with their gear. The casualty was then taken back to Gyllyngvase Beach where they were safely landed at 10.29am, the inshore lifeboat being released to return to its station.

The inshore lifeboat was recovered onto its slipway at 10.38am where it was refuelled and made ready for service by 11.00am.

The experienced windsurfer had been caught out when the wind had suddenly dropped leaving them unable to sail clear of the lee shore. They were extremely grateful for the assistance given.

DSC Distress Alert

Date: Wednesday 28 October 2020

Location: Falmouth Harbour, Penryn River and Carrick Roads

Weather: W 4, slight sea and swell, fair visibility, overcast with rain

All-weather lifeboat: (launched 4.01pm / returned 4.58pm) – Jonathon Blakeston (Coxswain), Carl Beardmore, Dave Nicoll, Sandy Procter, Adam West, Will Allen, Tamsin Mulcahy

Inshore lifeboat: (launched 3.59pm / returned 4.50pm) – James Wakefield (Helmsman), Tom Bird, Tamara Brookes, Lloyd Barron

Details:

At 3.39pm Falmouth Coastguard requested that the Falmouth lifeboat crew be brought to immediate readiness following the receipt of a VHF DSC distress alert which was believed to have come from a handheld radio. Following further enquiries at 3.55pm Falmouth Coastguard requested that both the all-weather and inshore lifeboats be launched to carry out a search of Falmouth Harbour, Penryn River and the Carrick Roads.

The inshore lifeboat launched from her slipway at 3.59pm with the all-weather lifeboat leaving its pontoon berth at 4.01pm. The Lifeboats were initially tasked with searching the Falmouth Inner Harbour and Penryn River. At 4.34pm all-weather lifeboat headed out to commence a search of the Carrick Roads while the inshore lifeboat continued its search of the River and Harbour. At 4.43pm it was confirmed that the source of DSC Alert had been located ashore so both lifeboats were released to their station.

The inshore lifeboat was back on its slipway by 4.50pm while the all-weather lifeboat was back alongside its pontoon berth at 4.58pm. Both lifeboats were refuelled and made ready for service by 5.10pm.

The DSC alert had been received on the coastguards Falmouth VHF Ariel but did not include a verbal distress message, so the nature and location of the distress was unknown. The source was confirmed to have been due to an equipment malfunction.

Multiple people in water from Capsized Dory

Date: Saturday 14 November 2020

Location: off Summers Beach, St Mawes

Weather: SW 6-7, moderate sea and swell, good visibility, partly cloudy

Inshore lifeboat: (launched 4.25pm / returned 4.59pm) – Neil Capper (Helmsman), Tom Bird, Tamara Brookes, Tamsin Mulcahy

Details:

At 4.17pm Falmouth Coastguard requested that Falmouth inshore lifeboat be launched following reports of six people in the water from a capsized Dory off Summers Beach, St Mawes. The Portscatho Coastguard Rescue Team was also tasked to assist.

The inshore lifeboat launched from her slipway at 4.25pm and arrived off St Mawes at 4.31pm. It was soon confirmed that the six occupants had managed to make the safety of the shore while their capsized dory had been washed up against the sea wall. The Coastguard Team checked that the occupants did not need any medical attention, they were found to be cold but not requiring any further assistance and they arranged their own transport to take them home. It was considered too risky to attempt to recover the Dory so at 4.47pm both the inshore lifeboat and Coastguard Team were released to return to their stations.

The inshore lifeboat was recovered onto its slipway at 4.59pm where it was refuelled and made ready for service by 5.15pm.

The six occupants of the Dory from Place Manor had gone across to St Mawes for essential goods and were on their way back when their boat got swamped leaving them in the water. It is believed that none of them were wearing lifejackets but thankfully they managed to make it to the shore despite the poor sea conditions. They were given safety advice by the Coastguard Team as the outcome of this incident could so easily have had a tragic ending.

Medical evacuation of injured person from rocks

Date: Thursday 26 November 2020

Location: Toll Point, Helford

Weather: NE 2, slight sea and swell, excellent visibility, partly cloudy

Inshore lifeboat: (launched 10.57am / returned 1.18pm) – Jamie Wakefield (Helmsman), Tamara Brookes, Cadan Harris, Lloyd Barron

Details:

At 10.48am Falmouth Coastguard requested that Falmouth inshore lifeboat be launched to assist in the medical evacuation of a person who had fallen on the rocks at Toll Point in the Helford. The Falmouth Coastguard Rescue Team and Ambulance Paramedics were also tasked to assist.

The inshore lifeboat launched from her slipway at 10.57am and having rounded Pendennis Point headed across Falmouth Bay arriving at Toll Point at the entrance to the Helford Estuary at 11.12am. Two casualty care trained lifeboat crew were put ashore to help assess and treat the casualty. Once the Coastguard Team and paramedics arrived on scene the casualty was given additional pain relief before being transferred to the inshore lifeboat on a stretcher. The casualty accompanied by a paramedic was taken to the beach at Durgan Village where they were transferred to a waiting ambulance at 12.55pm, the inshore lifeboat being released to return to its station.

The inshore lifeboat was recovered onto its slipway at 1.18pm where it was refuelled and made ready for service by 2.05pm.

The casualty had slipped on the rocks at Toll Point and was believed to have suffered a dislocated shoulder. Access from the shore side to the rocks was limited and the nearest vehicle access for the ambulance was a considerable distance away, so the best option was to transfer the casualty by lifeboat.

Open boat broken down

Date: Thursday 3 December 2020

Location: Gerrans Bay

Weather: NW 3, slight sea and swell, good visibility, overcast with rain

Inshore lifeboat: (launched 2.16pm / returned 3.50pm) – Nick Head (Helmsman), Tamara Brookes, Andy Edwards, Cadan Harris

Details:

At 2.06pm Falmouth Coastguard requested that Falmouth inshore lifeboat be launched to assist a 4.8m Open Boat with two persons on board which had dialled 999 to report that it had broken down between Portloe and Gull Rock.

The inshore lifeboat launched from her slipway at 2.16pm and having rounded Zone Point headed up to the East locating the casualty vessel off Killigerran Head at 2.27pm. A tow line was passed to the vessel and the tow commenced at 2.35pm. The vessel was towed back to Cowlands Creek in the River Fal where it was placed on its mooring off Roundwood Quay at 3.30pm, the inshore lifeboat being released to return to its station.

The inshore lifeboat was recovered onto its slipway at 3.50pm where it was refuelled and made ready for service by 4.10pm.

When the crew of the vessel contacted the coastguard, they advised that there were no other vessels in the area, so the inshore lifeboat was tasked to assist. The vessel did not have a VHF radio so had to raise the alarm using their mobile phone.

Motor Sailor broken down

Date: Thursday 31 December 2020

Location: Carrick Roads

Weather: NW 1, slight sea and no swell, fair visibility, overcast with sleet and snow

Inshore lifeboat: (launched 4.43pm / returned 4.53pm) – Nick Head (Helmsman), Tamsin Mulcahy, Andy Edwards, Joe Amps

Details:

At 4.30pm Falmouth Coastguard requested that Falmouth inshore lifeboat be launched to assist a 5.5m Motor Sailor which had broken down in the Carrick Roads in light airs.

The inshore lifeboat launched from her slipway at 4.43pm and headed out of the harbour, but at 4.46pm as the lifeboat was passing the docks, the coastguard confirmed that the boat had been taken in tow by another vessel, so the lifeboat was released to return to its station.

The inshore lifeboat was recovered onto its slipway at 4.53pm where it was refuelled and made ready for service by 5.10pm.

The occupants of the boat had contacted the coastguard to advise of their predicament but had also advised that they were hoping to obtain assistance from ashore. Due to the fading day light the coastguard decided to request the launch of the inshore lifeboat but as it passed the docks, they received confirmation from the coastguard that the vessel had been taken in tow and their assistance was no longer required.

How you can support Falmouth Lifeboat.

Thank you for your interest in Falmouth Lifeboat.

You can keep in touch by visiting our website and subscribing to our newsletter which will keep you up to date with all of the service calls.

You can also follow us on Facebook, Instagram, and Twitter. Please visit the web site for links etc.

How you can help.

See above and keep in touch. However, if you would like to donate to the RNL Falmouth Branch you can find the link on the website.

- You can share pages and content on the website to your friends.
- Share our Social Media posts to your friends.
- Purchase our Falmouth Lifeboat History book on the site.

If you are local and would like to volunteer to help fundraising, work in the shop, or as on board or shore based crew please get in touch.

Printed in Great Britain
by Amazon